AUTO
MATH
HANDBOOK

AUTO MATH HANDBOOK

Basic Calculations, Formulas, Equations and Theory for Automotive Enthusiasts

by

John Lawlor

HPBooks

HPBooks
A division of Penguin Group (USA) Inc.
375 Hudson Street
New York, New York 10014

© 1992, 1991 by John Lawlor

Printed in the U.S.A
34   33   32   31

Library of Congress Cataloging-in-Publication Data

John Lawlor.
Auto math handbook : mathematical calculations, theory,
and formulas for automotive enthusiasts / by John Lawlor.
   p. 160 cm.
  Includes biliographical references (p.) and index.
  ISBN 978-1-55788-020-8
  1. Automobiles—Mathematics.   I. Title
TL154.L38   1991   90-5088
529.2'0151—dc20   CIP

Contents

Acknowledgments

When I sat down to process the words that became the *Auto Math Handbook*, I had the support and guidance of several valued friends and fellow editors, writers and photographers, and I want to express my appreciation to them.

Past and present editors Duane Elliot of *Off-Road*, Jim McGowan of *Guide to Muscle Cars* and *Muscle Car Classics*, Spence Murray of *Mini-Truck*, Mike Parris of *Off-Road*, Ralph Poole of *Trailer Boats*, Leroi "Tex" Smith of *Hot Rod Mechanix* and the late Tom Senter of *Popular Hot Rodding* encouraged me to write about automotive mathematics and accepted articles I did on the subject from 1971 through 1990. Those articles were really the start of this book.

John Baker and the late Dick Cepek spurred my interest further by asking me to produce brief mathematical pieces for the catalogs issued by their respective firms, John Baker Performance Products of Webster, Wisconsin, and Dick Cepek, Inc. of Carson, California.

Former Petersen Publishing Co. librarian Jane Barrett and editors and writers Dean Batchelor, John Dinkel, Al Hall, Jon Jay and Jim Losee all suggested useful material for the book.

Friends Richard Shedenhelm and James J. Scanlan, M.D., read portions of the manuscript and offered excellent criticisms.

Photographer Chan Bush, who shot a portrait of me for my first book *How to Talk Car* in 1965, returned to perform the same service for *Auto Math Handbook*.

Gail Harrington, managing editor of *MotorHome Magazine*, was an unfailing source of personal encouragement and enthusiasm during the many months I had to spend my evenings and weekends working on the book.

Michael Lutfy, of HPBooks, had the unenviable task of editing the work of a fellow editor, and a highly opinionated one at that! He also went beyond the call of duty to supply many of the photos in the book.

Finally, there's Tom Madigan, a former editor of *Off-Road* and author of several books himself. He has given me his personal support, not only for this book but at several crucial points in my career. In gratitude, it is to Tom and his wife Darlene that *Auto Math Handbook* is dedicated.

Thank you one and all. Without your help, the job of writing the book would have been much more difficult, and it would not have turned out as well as it did.

About the Author

John Lawlor has been an automobile enthusiast since boyhood. He started writing about cars while still in college, with a weekly motoring column in the campus newspaper at Loyola University in Los Angeles, California.

He became a professional journalist in the late 1950s, when he went to work for Petersen Publishing Co., where he became a senior editor of *Motor Trend* and later, an editor in the firm's book division.

During the 1960s, he moved to Bond Publishing Company, where he was an assistant editor of *Car Life* and a contributor to the parent magazine, *Road & Track*. He has also served as the managing editor of *Popular Hot Rodding* and *Speed Age* magazines.

In 1967, he served as public relations director for the Inaugural Mexican 1000, the first off-road race down Mexico's Baja California peninsula. His efforts resulted in more event coverage by motor enthusiasts' magazines than any previous motorsport event of any kind. As a result, in 1979 Lawlor became the first, and as of this writing, only journalist or publicist elected to the Off-Road Hall of Fame, sponsored by the Specialty Equipment Market Association (SEMA). In 1989, he was similarly honored as one of the first ten inductees into the *Dune Buggies* and *Hot VWs Magazine* Hall of Fame.

He is the author of two books: *How to Talk Car*, a dictionary of automotive slang; and *Inside Full-Time Four-Wheel Drive*, a guide to the New Process 203 system published by Chrysler in the mid-70s.

John Lawlor is currently an editor for *Trailer Life* and *MotorHome* magazines. He lives in Studio City, California.

Photo by Chan Bush

Introduction

If you're seriously interested in automobiles and how they perform, sooner or later you'll have to deal with mathematics. Virtually all aspects of motorsports, from bore and stroke, through power and torque, to time and speed, involve mathematical calculations.

I recognized this as a young auto enthusiast myself in the 1950s, and I was pleased when I discovered a booklet called *Mechanics of Vehicles* by Jaroslav J. Taborek. It was a collection of 14 articles about the mathematics of motor vehicle behavior, originally published by *Machine Design* magazine in 1957.

It was the first publication I'd ever seen devoted to automotive mathematics as such. However, Taborek was a professional engineer and he wrote for his colleagues, not for enthusiasts like me. Much of his work was over my head and, in fact, some of it still is!

Then in 1961, an article called "Math and Formulas for Hot Rodders" by Don Francisco appeared in *Hot Rod Magazine Yearbook Number One*. It was only five pages long, but it provided some genuinely useful mathematics for the aspiring hot rodder, and none of it required more than a grade school background in math. To the best of my knowledge, it was the first such compilation especially for car enthusiasts.

In the years since, I've seen numerous magazine articles and book chapters dealing with various aspects of auto math, but they've all been at one or the other of the extremes represented by Taborek's and Francisco's pioneering and long out-of-print efforts. They've been either ponderous professional tomes or frankly sketchy popular works. There's been no book-length collection of practical, elementary math for auto enthusiasts of average

education. That's a gap I've tried to fill with this work.

I've concentrated on math of genuine interest to the enthusiast, and avoided anything too specialized. That's dictated a particular emphasis on the engine and drivetrain, which are the core of true hot rodding.

I haven't included formulas that involve advanced forms of math. I've tried to write to a high school level and, in the pages that follow, there's plenty of arithmetic and algebra and even a little geometry, but there's no calculus. If you just breathed a sigh of relief, so did I!

I've also avoided most formulas that would require data simply not available to those who aren't working automotive engineers, such as the math used in aerodynamics or suspension design.

However, I have included the formulas for horsepower and torque, despite the fact that their measurement requires equipment not likely to be found in the average hot rodder's garage (though a dynamometer is likely to be more accessible than a wind tunnel or a suspension testing facility).

The interrelationships of horsepower and torque are among the most important principles of automotive engineering, and can't be ignored in a book devoted to automotive mathematics.

As a result, I believe the work serves as a useful primer of auto engineering and performance fundamentals, as well as a handbook of auto math. I hope it will have particular appeal to younger enthusiasts who are just developing an interest in the technology of auto performance, as I was when I first came across Taborek's and Francisco's works.

Many of the formulas I've presented could be worked on a simple 4-function arithmetic calculator. However, some of them will be much easier on an inexpensive scientific calculator, with *pi* and parentheses keys, and a few require a calculator which can find either square or cube roots.

The problem examples in the text were worked out to eight digits, because that is the capacity of most inexpensive calculators. However, the solutions were generally rounded off to no more than three decimal places and sometimes to none at all, depending on the degree of precision that seemed appropriate in each case.

In the text, single-digit whole numbers or integers are followed by a decimal point and a zero, e.g., five is 5.0; numbers less than one—or 1.0—have a zero preceding the decimal point, e.g., five/tenths is 0.5.

To enhance the value of the book as a reference and make it simpler to look up a specific formula, each chapter concludes with a table summarizing the formulas it has covered. Further, most of the formulas are written in plain English or easily recognized abbreviations such as *rpm* and *mph*, rather than in algebraic symbols, to make them as clear as possible to the non-mathematician.

A working engineer who happens to see this volume may criticize the limited attention given to the metric system of measurement—or to identify it more properly, the *Systeme International des Unites*, or S.I. for short.

The professional may work with such S.I. units as the meter for length, kilogram for weight and watt for power. However, the kind of enthusiast I had in mind as I wrote this book continues to measure in traditional feet, pounds and horsepower, not in their metric or S.I. equivalents.

I've tried to be as complete and accurate as possible in the material I've presented. If there are any omissions or errors, I'd appreciate having them brought to my attention so that the necessary additions or corrections can be made in future editions of this book.

Displacement, Stroke & Bore 1

A clear understanding of the relationship between displacement, stroke and bore is essential for high performance engine building (photo by Michael Lutfy).

The *displacement* or cubic capacity of an automobile engine is the volume within the cylinders *swept* by the pistons as each of them makes one full stroke downward, from top dead center to bottom dead center, or one stroke upward, from bottom to top. However, this swept volume is not the total volume of the cylinder. The total includes not only the swept volume but also the combustion chamber volume above the piston at top dead center.

PISTON DISPLACEMENT

To find an engine's overall swept volume, or *piston displacement*, as it's also called, you must know the engine's *bore* and *stroke*. As measurements, the stroke is the distance the piston travels downward or upward in the cylinder and the bore is the diameter of the cylinder.

If you have the factory figures for the bore and stroke, you would probably have the displacement too. So why do

1

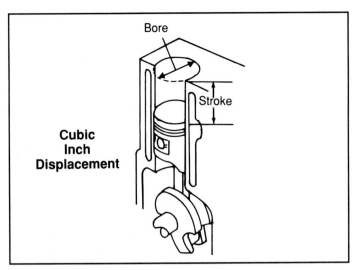

Bore

Stroke

**Cubic
Inch
Displacement**

The swept volume in cubic inches of an individual cylinder is found by multiplying pi/4 by the bore in inches squared by the stroke in inches. The overall swept volume of an engine is simply the volume of one cylinder multiplied by the total number of cylinders.

you need a formula for finding it? One reason is that the displacement given in the factory-supplied spec sheets is usually rounded up or down to a whole number or integer by the factory. When it comes to precise engine building, it can be instructive to double-check the stock specs and find the exact figure.

More importantly, if you're building a racing engine to compete in a class with a specific displacement limit and you want to increase or decrease the displacement by modifying the stroke or bore, you need to know how to calculate the effects of particular modifications. If you change the stroke or bore, how much will the displacement change?

To find the overall engine displacement, you must first find the swept volume of a single cylinder, and then multiply that figure by the number of cylinders. For that, you need to know the formula for the volume of a cylinder as a geometric shape, which is *pi*/4 or 3.1415927/4, which equals 0.7853982. This number is then multiplied by the square of the diameter and by the height. In an engine, the diameter of a cylinder is the bore, and the height is the stroke, so the formula for finding the cylinder volume in cubic inches is; *pi* divided by 4, multiplied by the bore squared in inches multiplied by the stroke in inches or:

Cylinder Volume = $pi/4 \times bore^2 \times stroke$

To find the overall piston displacement of the entire engine, you'd take the value of this equation and multiply that by the number of cylinders in the engine.

Example—As an example, let's see what the displacement would be for an 8-cylinder engine with a 4.0-inch bore and 3.48-inch stroke:

Displacement = $0.7853982 \times 4^2 \times 3.48 \times 8$

The answer is 349.84776 cubic inches, which can be rounded up to an even 350, because the decimal figure is well over 0.5. Those happen to be the measurements of a version of the Chevrolet small-block V-8.

Decimal Conversion—If either the bore or stroke includes a common fraction, it must be converted to a decimal before being entered on an arithmetic calculator. To demonstrate, take an 8-cylinder engine with a bore of 3-7/8 inches and a stroke of 3-1/4 inches. In decimals, those figures would be 3.875 and 3.25 respectively, and would plug into the formula this way:

Displacement = $0.7853982 \times 3.875^2 \times 3.25 \times 8$

Here, the displacement is 306.62435 cubic inches or, rounded up, 307. This, too, is a variation of the Chevy small-block V-8.

On a scientific calculator with parentheses keys, it wouldn't be necessary to prefigure the decimal equivalents of the fractions. They could be entered as follows:

Displacement = $0.7853982 \times (3+7/8)^2 \times (3+1/4) \times 8$

The answer is again 306.62435. But note that plus [+] signs must be used between the integers and fractions.

Rounding Up or Down—I mentioned earlier that it can be instructive to double-check factory displacement figures. As a case in point, from 1958 through 1966, Ford built a V-8 engine with a bore of 4.0 inches and a stroke of 3.5 inches, and advertised its displacement as 352 cubic inches. Then, in 1969, the company introduced a new, lighter-weight V-8 with the same bore and stroke but, this time, claimed a displacement of 351 cubic inches. To find which figure is closer to the truth, try the formula with a 4.0-inch bore and 3.5-inch stroke:

Displacement = $0.7853982 \times 4^2 \times 3.5 \times 8$

The actual displacement is 351.85838 cubic inches, which should be rounded up to 352 because the decimal is well over 0.5. However, Ford has chosen to round it *down* to 351 for the later, better-performing engine, probably to avoid confusing it with the older, less efficient unit.

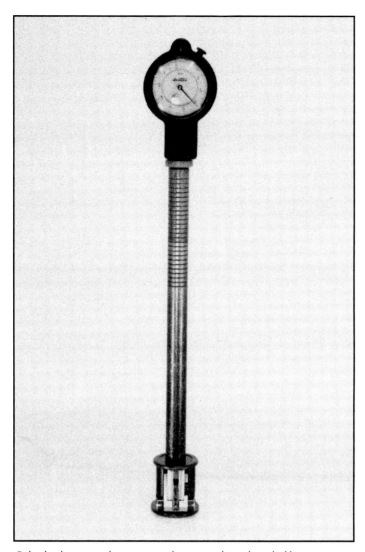

Cylinder bore can be measured accurately with a dial-bore gauge. This is an essential tool for those who want to do their own boring or honing (photo by Larry Shepard).

The bore should be checked carefully at the top, center and bottom of piston travel and in two different directions horizontally at each level (photo by Larry Shepard).

Overboring—Staying for the moment with the earlier Ford 351—uh, 352—imagine that you have a well-worn specimen that needs a "clean-up" overbore of .030-inch. How would that affect the displacement? To find out, try the formula again with the bore increased to 4.03 inches:

$$\text{Displacement} = 0.7853982 \times 4.03^2 \times 3.5 \times 8$$

The answer is 357.15604 cubic inches or, rounded down, 357 cubic inches.

Now, what if you wanted to know how much you could modify either the stroke or the bore, yet stay within a specific cubic-inch limit?

▼A micrometer and snap gauge can also be used to measure bore, but this combination of instruments isn't usually as reliable as the dial-bore gauge (photo by Larry Shepard).

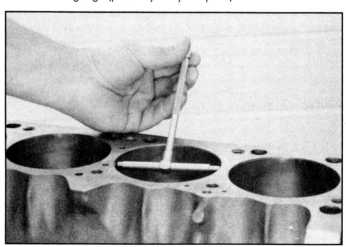

STROKE

The formula for the stroke is the displacement, divided by one-fourth of *pi*, multiplied by the square of the bore, multiplied by the number of cylinders:

$$\text{Stroke} = \frac{\text{displacement}}{(pi/4 \times \text{bore}^2 \times \text{no. of cylinders})}$$

Example—Suppose you have a car with a late-model Ford 352—uh, 351—V-8 and you want to race in a class with a limit of 366 cubic inches. You want to keep the stock 4.0-inch bore but stretch the 3.5-inch stroke. How far can you go with it?

When boring or honing cylinders, a steel honing plate approximately 1.0-inch thick should be bolted to the top of the block to simulate cylinder head stress (photo by Larry Shepard).

The displacement you are concerned with is 366, the bore 4.0 inches and the number of cylinders is, of course, 8:

$$\text{Stroke} = \frac{366}{(0.7853982 \times 4^2 \times 8)}$$

The maximum allowable stroke within the 366 cubic-inch limit would be 3.6406693 inches or, rounded down, 3.64 inches, 0.14-inches more than stock.

BORE

To find either the displacement or the stroke, you have been using the square of the bore. Conversely, to find the bore, you will have to work with the *square root* of the other factors. The formula for the bore is the square root of the displacement, divided by one-fourth of *pi*, multiplied by the stroke, multiplied by the number of cylinders or:

$$\text{Bore} = \sqrt{\frac{\text{displacement}}{(pi/4 \times \text{stroke} \times \text{no. of cylinders})}}$$

Example—If, when building the Ford V-8 for that 366-cubic-inch limit described earlier, you decided to stick with the 3.5-inch stroke, how big a bore could you use?

$$\text{Bore} = \sqrt{\frac{366}{(0.7853982 \times 3.5 \times 8)}}$$

The answer is 4.0795906 inches. That is the absolute maximum, however, so do not round it up to 4.08 inches, even though the decimal is over 0.5, or you will be over the limit dictated by the rule book! To demonstrate that point, try a bore of 4.08 and stroke of 3.5 in the formula for displacement:

$$\text{Displacement} = 0.7853982 \times 4.08^2 \times 3.5 \times 8$$

Those figures provide 366.07346 cubic inches—and an engine that is too big for a 366 cubic-inch class. As the National Hot Rod Association states in its drag race rules, "Any part of a cubic inch is rounded off to the next highest inch. . . ." In other words, NHRA officials would consider your overbored engine to have 367 cubic inches, and would disqualify it. So take advantage of rounding up when you can. But, like Ford with its 352—uh, 351—know when to round down, too.

The Sunnen CK-10 is a popular type of honing machine found in most serious engine-building shops. In the hands of a skilled operator, it's extremely accurate (photo by Larry Shepard).

The cylinders in this engine have been bored and honed, using a Sunnen CK-10 and a honing plated bolted to the top of the block (photo by Michael Lutfy).

1970 through 1973. It had a bore of 83 millimeters and a stroke of 73.7 millimeters, or, in the formula:

$$\text{Displacement} = \frac{0.7853982 \times 83^2 \times 73.7 \times 6}{1000}$$

That provides a displacement of 2392.5708 or, rounded up, 2393 cubic centimeters.

Converting—However, a simpler, more direct way to find the displacement in cubic centimeters would be to convert the bore and stroke to centimeters by dividing them by 10 before entering them. In the 240Z, that would change them from 83 to 8.3 and from 73.7 to 7.37:

$$\text{Displacement} = 0.7853982 \times 8.3^2 \times 7.37 \times 6$$

The result is again 2392.5708 or 2393 cubic centimeters. In using the formulas for bore and stroke, you can work with the figures in centimeters and multiply the result by 10 to convert to millimeters.

To continue with the Z car: In 1974, the 240Z became the 260Z, with the engine enlarged to 2565 cubic centimeters. The bore was still 83 millimeters or 8.3 centimeters, but the stroke had been increased. You can find the new stroke in centimeters by entering the displacement in cubic centimeters and the bore in centimeters:

$$\text{Stroke} = \frac{2565}{(0.7853982 \times 8.3^2 \times 6)}$$

METRIC DISPLACEMENT

In the metric system, the bore and stroke are usually given in millimeters and the displacement in cubic centimeters. If you use the displacement formula as is, and enter the bore and stroke in millimeters, the result will be in cubic millimeters, which must be divided by 1000 to be converted to cubic centimeters. You can accomplish that by changing the formula to:

$$\text{Displacement (cc)} = \frac{pi/4 \times bore^2 \times stroke \times no.\,of\,cylinders}{1000}$$

Example—Let's try this with the 6-cylinder engine in Nissan's original Z car, the Datsun 240Z, offered from

According to the NHRA, "Any part of a cubic inch is rounded off to the next highest inch..." so be careful how you round off your figures. A few thousandths over could mean disqualification (photo by Michael Lutfy).

The result is 7.9011454 centimeters or, multiplied by 10 and rounded down, 79 millimeters.

For 1975, the 260Z evolved into the 280Z, with the engine displacing 2753 cubic centimeters. The stroke was the same 79 millimeters or 7.9 centimeters as in the 260Z, but the bore had been enlarged. Here is the calculation to find the new cylinder dimension:

$$\text{Bore} = \sqrt{\frac{2753}{(0.7853982 \times 7.9 \times 6)}}$$

The bore was 8.5994167 centimeters or, multiplied by 10 and rounded up, an even 86 millimeters.

Centiliters—If you wonder about the significance of those figures—240, 260 and 280—they represent the approximate engine displacement in centiliters, a centiliter being 1/100 of a liter. A liter, in turn, is 1000 cubic centimeters.

Mercedes-Benz is another make which uses centiliters of engine displacement to designate specific models. The Mercedes 300, for example, has a 3.0-liter engine and the 500 a 5.0-liter.

Finally, there's the question of converting back and forth between the two systems of measurement, with inches and cubic inches on the one hand, and millimeters, cubic centimeters and liters on the other. Factors for these conversions and many others will be found in Appendix A, beginning on page 123.

Table 1

FORMULAS FOR DISPLACEMENT, STROKE & BORE

$pi = 3.1415927$

$pi/4 = 0.7853982$

cylinder volume $= pi/4$ x bore^2 x stroke

displacement $= pi/4$ x bore^2 x stroke x number of cylinders

$$\text{stroke} = \frac{\text{displacement}}{(pi/4 \times \text{bore}^2 \times \text{number of cylinders})}$$

$$\text{bore} = \sqrt{\frac{\text{displacement}}{(pi/4 \times \text{stroke} \times \text{number of cylinders})}}$$

Compression Ratio 2

The higher the compression ratio, the greater the combustion, which results in greater power. That's the credo most drag racers live by (photo by Jim Kelso).

The compression ratio of an engine is the relationship between the combined cylinder and combustion chamber volumes with the piston at bottom dead center and the combustion chamber volume with the piston at top dead center.

The higher the compression ratio is, the more the air/fuel mixture will be compressed. And the more the mixture is compressed, the more powerful combustion will be. However, as the mixture is compressed, it gets hotter and there is a danger that some of it may ignite prematurely. That phenomenon is called *detonation* or *knock*.

The octane rating of a gasoline is a measure of its resistance to knock. The higher the octane is, the greater the resistance will be. And the greater the resistance to

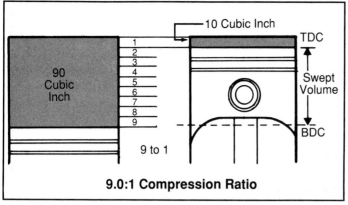

9.0:1 Compression Ratio

The compression ratio is the relationship between the combined volume of the cylinder and combustion chamber with the piston at bottom dead center and the volume of the combustion chamber with the piston at top dead center.

knock is, the higher the compression ratio can be. So that's why high-powered engines have high-compression ratios and use high-octane fuels.

CALCULATING COMPRESSION RATIO

The compression ratio isn't difficult to calculate; it's equal to the sum of the cylinder volume and combustion chamber volume divided by the latter, or:

$$\text{Compression Ratio} = \frac{\text{cylinder + chamber volume}}{\text{chamber volume}}$$

The cylinder volume, as explained in the previous chapter, can be found with the formula:

$$\text{Cylinder Volume} = pi/4 \times \text{bore}^2 \times \text{stroke}$$

There's one slight catch: Combustion chamber volume is usually measured in cubic centimeters or cc's, so you'll also want the cylinder volume in cc's. That means entering the bore and stroke into the formula in centimeters.

If you already have the bore and stroke in inches, you simply multiply by 2.54 to convert to centimeters. If, on the other hand, you have the cylinder dimensions in millimeters, you only need to divide by 10 to find what they are in centimeters.

Example—Let's use the 4.0-inch bore and 3.5-inch stroke of the Ford 351 V-8 as an example. Multiplied by the conversion factor of 2.54, those figures would become 10.16 and 8.89 centimeters respectively, and plug into the formula thusly:

$$\text{Cylinder Volume} = 0.7853982 \times 10.16^2 \times 8.89$$

That indicates a cylinder volume of 720.74072 cubic centimeters.

MEASURING CHAMBER VOLUME

Now comes the hard part: Finding the combustion chamber volume or, as it's also known, the *clearance* or *compression* volume. Because the combustion chamber is irregular in shape, its volume cannot be calculated with a simple formula. You will have to measure it physically and, to do that, you will need a burette marked in cc's and filled with a light oil, cleaning solvent or even water. The procedure is called, logically enough, *cc-ing*.

Robert Landy of drag racer Dick Landy's DLI engine building business in Northridge, California, zeros out the burette as he prepares to cc a cylinder head of a Chrysler small-block V-8 engine (photo by Michael Lutfy).

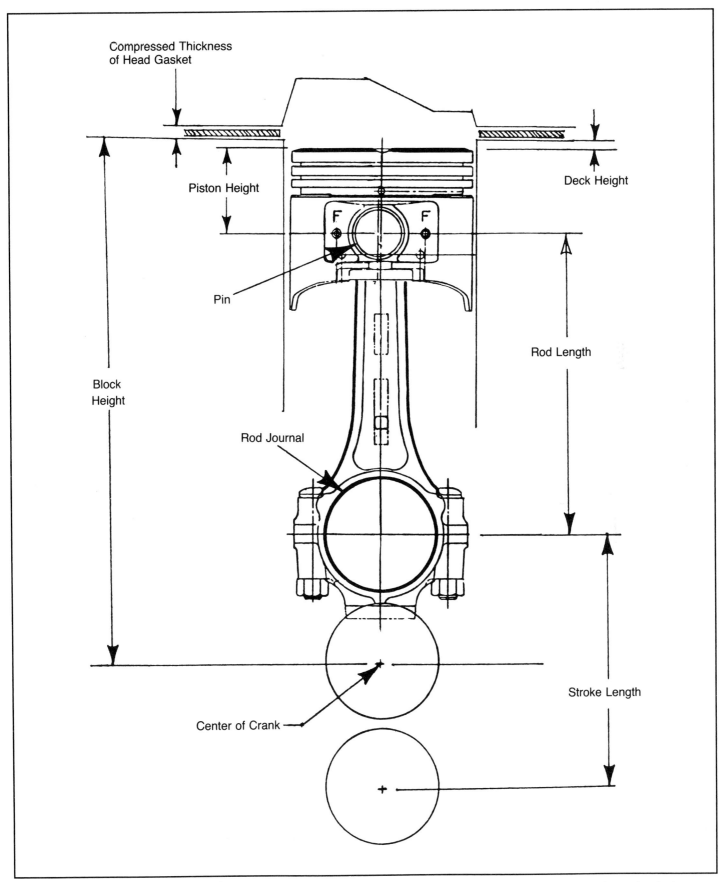

This illustration provided by Larry Shepard of Mopar Performance shows rod, piston and cylinder dimensions. Note the gasket thickness and deck height at the top. Slight as these may be, they affect measurement of the combustion chamber volume.

11

$$\text{Compression Ratio} = \frac{(720.74072 + 92.5)}{92.5}$$

The compression ratio would be 8.7917915 or, rounded up, 8.8:1.

CC-ing with Engine Disassembled—If the engine is disassembled, measuring the combustion chamber volume is considerably more difficult. First, you have to cc the head. Then you have to calculate the added volume that will be provided by the gasket. If the engine had flat-topped pistons that, at top dead center, were perfectly even with the deck, or top of the block, that's all you'd need.

However, that isn't true of many engines. At top dead center, the piston may stop short of the deck height. In addition, if the piston top is dished, or concave, it will increase the combustion chamber volume; if it's domed, or convex, it will decrease the volume.

1/2-Inch Downfill Method—In HPBooks' *How to Hot Rod Small-Block Mopar Engines*, Larry Shepard of Chrysler's *Mopar Performance* offers a technique for cc-ing the block he calls the "0.500-inch downfill" or "1/2-inch downfill" method. With this method, the head is removed and the engine is positioned so that the cylinder to be measured is vertical. Then the piston is lowered 1/2 inch or 1.27 centimeters from top dead center. The distance is arbitrary. The point is simply to be sure that the entire piston, including the dome, is below the deck and fully within the block.

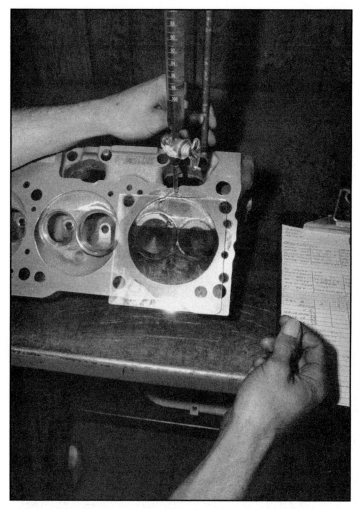

A flat, 1/4-inch plate of clear Plexiglass is placed over the combustion chamber, so the fluid used to measure chamber capacity can be seen clearly. Once the chamber is full, the burette is checked and the amount of fluid used is recorded (photo by Michael Lutfy).

CC-ing With Engine Assembled—With the engine fully assembled and mounted on a stand, it should be tilted so the spark plug hole in the cylinder to be measured is vertical. With the piston at top dead center, the valves closed and the spark plug removed, pour liquid from the burette through the plug hole until it reaches the beginning of the plug threads. The amount poured from the burette will indicate the combustion chamber volume.

Example—Suppose you cc just one cylinder head in the Ford 351 engine I've been using as an example and find that the combustion chamber volume is 92.5 cubic centimeters, and apply that in the formula for compression ratio:

It's not unusual for normally aspirated race engines to run as high as an 11-to-1 compression ratio (photo by Michael Lutfy).

Now, with the burette, you can find the volume above the piston. By using the formula for cylinder volume, you can find what the volume would be if the piston were flat-topped. The difference between that figure and the volume you find by cc-ing tells you how much a dished piston increases the overall combustion chamber volume or how much a domed piston decreases it.

Adding Gasket Thickness—Okay, getting back to your Ford engine, suppose you cc the portion of a combustion chamber in one of the heads and find it has a volume of 75 cc's. Your gasket is 0.045-inch or 0.1143-centimeters thick; to find how much that will add to the chamber volume, apply the formula for cylinder volume using the gasket thickness in place of the stroke:

$$\text{Volume} = 0.7853982 \times 10.16^2 \times 0.1143$$

That works out to 9.2666664 cc. With flat-topped pistons even with the deck at top dead center, your total combustion chamber volume would be 75 plus 9.2666664 or 84.266667 cc or, rounded up, 84.27 cc.

What would that give you in the way of compression ratio? To find out, add the cylinder volume to the chamber volume, and divide by the chamber volume, or:

$$\text{Compression Ratio} = \frac{(720.74072 + 84.266667)}{84.266667}$$

The answer is 9.5526606 or, rounded down, 9.55:1.

This instrument is used to measure the deck height, i.e. the distance between the top of the piston and the top of the block (photo by Larry Shepard).

CALCULATING CHAMBER VOLUME

If you can use combustion chamber volume to find compression ratio, can you do the opposite and use compression ratio to find chamber volume? Indeed, you can, and it's a useful thing to know how to do in order to find exactly what chamber volume you need for a specific compression ratio.

To find combustion chamber volume, divide the cylinder volume by the compression ratio minus 1.0, or:

$$\text{Chamber Volume} = \frac{\text{cylinder volume}}{\text{compression ratio} - 1.0}$$

Example—If you wanted to increase the compression ratio of your Ford 351 to 10.5:1, what would the overall combustion chamber volume have to be?

$$\text{Chamber Volume} = \frac{720.74072}{(10.5 - 1.0)} \quad \text{or} \quad \frac{720.74072}{9.5}$$

The domes of pistons can vary greatly in shape, complicating the task of determining combustion chamber volume. One method of measuring the dome volume is with the 1/2-inch downfill technique described in the text (courtesy of Childs & Albert).

The answer would be 75.867444 or, rounded up, 75.87 cc. You can double-check those results by reverting to the formula for compression ratio:

$$\text{Compression Ratio} = \frac{(720.74072 + 75.867444)}{75.867444}$$

Sure enough, that provides a compression ratio of 10.5:1, bringing you right back to where you started.

MILLED HEADS

One of the most popular ways to increase compression ratio is to *mill* the cylinder heads. The question is, how much do you mill the heads for a specific increase in compression?

Displacement Ratio—The formula for determining that is based on the *displacement ratio*, which is cylinder volume divided by combustion chamber volume, and it is always 1.0 less than the compression ratio. For example, the displacement ratio for a cylinder with a 9.55:1 compression ratio is 8.55:1.

Example—Suppose you want to raise the compression ratio from 9.55 to 11:1. How much do you mill the heads?

The existing displacement ratio is, as you've seen, 8.55, while the new displacement ratio will be 11 minus 1.0 or 10. Begin by subtracting the old ratio from the new and dividing the result by the product of the two ratios multiplied together. That figure is then multiplied by the stroke in order to find the amount the heads should be milled. Here's all that expressed as an equation:

Milling the heads is a common techniques used to increase compression ratio. The question is, how much do you mill to gain a specific increase in compression ratio? The formula for determining that is based on the displacement ratio, which is cylinder volume divided by combustion chamber volume, and it is always 1.0 less than the compression ratio. For details, see the text (photo by Michael Lutfy).

$$\text{Amount To Mill} = \frac{\text{new disp. ratio - old disp. ratio}}{\text{new disp. ratio} \times \text{old disp. ratio}} \times \text{stroke}$$

Milling in Inches—If the stroke were 3.5 inches, the figures plugged into the formula would look like:

$$\text{Amount To Mill} = \frac{10 - 8.55}{10 \times 8.55} \times 3.5 = \frac{1.45}{85.5} \times 3.5$$

That gives you 0.0169591 times 3.5 or 0.0593567 inch or, rounded up, 0.060 inch to mill the heads for our desired increase from 8.55:1 to 11:1.

Milling in Millimeters—The formula will also work if you have the stroke in millimeters. Multiplied by a conversion factor of 25.4, a stroke of 3.5 inches would be 88.9 millimeters. To find how much the heads should be milled in millimeters for the same increase in compression ratio:

$$\text{Amount To Mill} = \frac{10 - 8.55}{10 \times 8.55} \times 88.9 = \frac{1.45}{85.5} \times 88.9$$

That works out to 0.0169591 x 88.9 or 1.5076608 or, rounded up, 1.51 millimeters to mill the heads in order to increase the compression ratio from 8.55:1 to 11:1.

Now, if you've been observant, you'll notice that if you calculated the amount to mill in inches first, you wouldn't necessarily need to recompute the entire formula again to know what it would be in millimeters. All you have to do is multiply the answer in inches, 0.0593567, and multiply it by the conversion factor, 25.4. The result is 1.5076608, exactly the same as the answer in millimeters.

Robert Landy gets set to mill the surface of a cylinder head. Let's hope he has made the proper calculations. Once metal is shaved off, it can't be put back!

Table 2

FORMULAS FOR COMPRESSION RATIO

$$\text{compression ratio} = \frac{\text{cylinder + chamber volume}}{\text{chamber volume}}$$

$$\text{cylinder volume} = pi/4 \times \text{bore}^2 \times \text{stroke}$$

$$\text{chamber volume} = \frac{\text{cylinder volume}}{\text{compression ratio - 1.0}}$$

$$\text{displacement ratio} = \frac{\text{cylinder volume}}{\text{chamber volume}}$$

$$\text{amount to mill} = \frac{\text{new disp. ratio - old disp. ratio}}{\text{new disp. ratio} \times \text{old disp. ratio}} \times \text{stroke}$$

Piston Speed 3

Drag racers need to be especially concerned with piston speed. If it gets too high, the piston will outrun flame front travel and lose power at best. At worst, a rod or piston will break (photo by Michael Lutfy).

Piston speed is the rate at which the piston travels up and down in the cylinder, and is usually measured in feet per minute. The rate isn't constant. At higher rpm, the piston may be going more than 100 miles per hour near the middle of its stroke. It slows as it approaches either end of its cycle, comes to a momentary stop as it hits top or bottom dead center, and then accelerates as it starts back in the other direction.

In other words, it may be going from zero to over 100 and back to zero during each stroke—over a distance of only 2.0, 3.0, or 4.0 inches!

If the piston speed gets too high, the primary hazard is that a piston or rod—or more than one of them—may break

from the strain. Therefore, determining piston speed will also help determine a practical rpm limit.

A secondary problem is that a piston may outrun *flame front travel*—running faster than the expanding air/fuel mixture is pushing it—and effective pressure and, with it, horsepower will drop. That's not as serious as breakage, but it's not exactly desirable either.

It's possible to determine the exact piston speed, as well as the rate of acceleration or deceleration, at any point during the cycle, but it takes differential calculus to do it. Fortunately, you don't have to worry about that. All you need to find is the average or mean piston speed, and that can be done with a relatively simple formula.

MEAN PISTON SPEED

A piston makes two full strokes, one up and one down, during each crankshaft revolution. Therefore, the mean piston speed in inches per minute would be two times the stroke in inches, times the crankshaft revolutions per minute, or rpm. The result is divided by 12 to convert it to feet per minute or fpm, and the formula is:

$$\text{Piston Speed in fpm} = \frac{2 \times \text{stroke in inches} \times \text{rpm}}{12}$$

If you divide both the numerator and denominator in the equation by 2, you can reduce that to:

$$\text{Piston Speed in fpm} = \frac{\text{stroke in inches} \times \text{rpm}}{6}$$

In that form, it'll obviously be a little easier to work with.

Example—In the early days of hot rodding, when the flathead Ford V-8 reigned supreme at the dry lakes and dirt tracks, 2500 feet per minute was considered the maximum practical piston speed—not just for Fords, but for all cars. How did the flathead Ford stack up against that norm of 2500 feet per minute? Introduced in 1932, the early Ford V-8 had a displacement of 221 cubic inches, with a bore of 3.06 inches and, more to our immediate point, a stroke of 3.75 inches. Let's see what its piston speed was at 4000 rpm:

$$\text{Piston Speed} = \frac{3.75 \times 4000}{6}$$

That put it right at 2500 feet per minute! Flatheads may have been revved beyond 4000 rpm, but not for very long, at least in stock form!

In 1962, just 30 years after the debut of the flathead, Ford introduced another 221 cubic-inch V-8. Called the Fairlane V-8 after the mid-size series where it was first used, it was the forerunner of Ford's modern small-block V-8s and featured a relatively big bore and short stroke, with cylinder dimensions of 3.50 by 2.87 inches. And what was its piston speed at 4000 rpm?

$$\text{Piston Speed} = \frac{2.87 \times 4000}{6}$$

That works out to only 1913 feet per minute, well under the traditional maximum of 2500.

With modern advances in metallurgy, higher piston speeds have become possible, but there are still limits. British tech author A. Graham Bell in his book *Performance Tuning in Theory and Practice* suggests what some of the limits are. With a stock cast-iron crankshaft and connecting rods, he recommends a maximum of 3500 feet per minute and, with a forged crank and heavy-duty rods and main bearing caps, between 3800 and 4000 feet per minute. At the outer limit, he believes that an all-out drag racing engine which is equipped with super-duty components, run on racing fuel with fast flame front travel and revved to the max for only a few seconds at a time, may get away with piston speeds as high as 5000 to 6000 feet per minute.

REV LIMITS

What if you wanted to know how high you could rev an engine without exceeding a specified piston speed? The formula is:

$$\text{RPM} = \frac{\text{piston speed in fpm} \times 6}{\text{stroke in inches}}$$

Let's see what the engine speed of the Fairlane V-8 with its stroke of 2.87 inches would've been at the old piston speed limit of 2500 feet per minute:

$$\text{RPM} = \frac{2500 \times 6}{2.87}$$

It could've been run all the way up to 5226 rpm. And what of its potential against the modern piston speed standard for stock engines of 3500 feet per minute?

$$\text{RPM} = \frac{3500 \times 6}{2.87}$$

That works out to 7317 rpm.

Our comparison of the flathead and Fairlane V-8s is instructive because it demonstrates that the same displacement is possible with radically different strokes. Let's be honest, though, 221 cubic inches was small for a V-8, even by today's slimmed-down standards, and the early Fairlane's 2.89-inch stroke was unusually short. It shouldn't have been surprising that the engine was capable of high rpm without excessive piston speed.

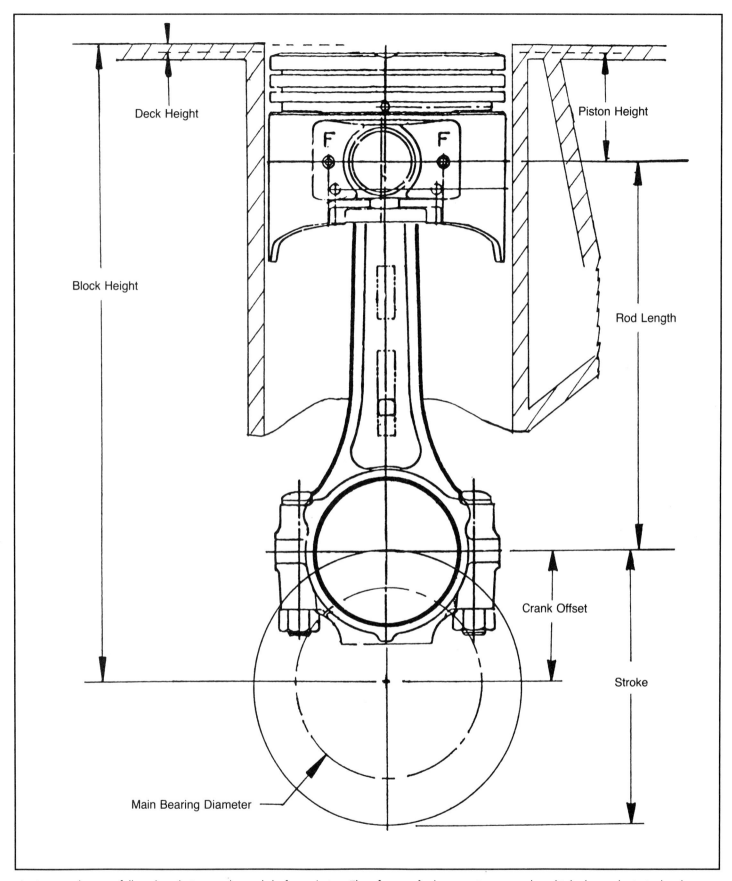

A piston makes two full strokes during each crankshaft revolution. Therefore, to find mean piston speed, multiply the stroke in inches by two, then multiply that figure by the rpm. Divide by 12 to find the piston speed in feet per minute (fpm). Courtesy Larry Shepard.

SMALL-BLOCK V-8 PISTON SPEEDS

But what of high-performance, small-block V-8s of the kind found in Mustangs and Thunderbirds? Or, for that matter, in Camaros and Corvettes?

Among the derivatives of that original Fairlane engine are V-8s of 302 and 351—uh, 352—cubic inches. Chevrolet has had comparable engines of 302 and 350 cubic inches.

302s—The 302s were developed in the late 1960s to fit the 5.0-liter limit in the Sports Car Club of America's Trans-Am series. Both Ford and Chevrolet used the same cylinder dimensions, a bore of 4.0 inches and stroke of 3.0 inches. Therefore, at any given rpm, they had the same piston speed. For example, at 8000 rpm, the figures would be:

$$\text{Piston Speed} = \frac{3 \times 8000}{6}$$

Or 4000 feet per minute.

350/351—Bores of 4.0 inches are also used in the Ford 351 and Chevy 350, but the strokes are 3.5 and 3.48 inches, respectively so, at any given rpm, the piston speeds will be similar but not the same.

For example, at 8000 rpm, the figures for the Ford 351 would be:

$$\text{Piston Speed} = \frac{3.5 \times 8000}{6}$$

Which would be 4667 feet per minute.

For the Chevy 350:

$$\text{Piston Speed} = \frac{3.48 \times 8000}{6}$$

And that works out to 4640 feet per minute.

In either case, you'd be pressing your luck unless you used the highest quality internal parts possible.

Drag Racing—For drag racing, you'll occasionally hear of Chevy small-block V-8s which are run from 10- to 12,000 rpm. However, these have been destroked, generally to 290 cubic inches, so to determine the stroke, apply the formula for stroke discussed in Chapter 1:

$$\text{Stroke} = \frac{290}{(pi/4 \times 4^2 \times 8)}$$

You'll find that a 290-cubic-inch V-8 with a 4.0-inch bore would have a stroke of 2.88 inches. Therefore, at 10,000 rpm:

$$\text{Piston Speed} = \frac{2.88 \times 10,000}{6}$$

The piston speed would be 4808 feet per minute.

At 12,000 rpm, the piston speed would rise to 5760 feet per minute, and that's about as far as anybody ought to go!

Between modern short-stroke engine designs and ongoing improvements in metallurgy, the recommended maximums in piston speed have become so high that some hot rodders don't pay much attention to them any more. But it's still wise to be aware of them, because there is a point at which even the best bearings and rods can fail, particularly when a powerplant is run consistently at higher-than-average rpm.

Table 3

FORMULAS FOR PISTON SPEED

$$\text{piston speed in fpm} = \frac{\text{stroke in inches} \times \text{rpm}}{6}$$

$$\text{rpm} = \frac{\text{piston speed in fpm} \times 6}{\text{stroke in inches}}$$

Brake Horsepower & Torque 4

Horsepower is defined as the measure of the ability to move a given weight a given distance in a given period of time. There couldn't be a more direct or dramatic example of what that really means than the performance of a Top Fuel dragster. This one, campaigned by Kenny Bernstein of Orange, California, will propel its estimated 1200 pounds from a standing start to the end of a 1/4 mile in slightly less than 5.0 seconds (photo by Michael Lutfy).

HORSEPOWER

Horsepower is the measure of the ability to move a given weight a given distance—that is, to apply leverage—in a given period of time.

Concept—The concept dates back to the 17th Century and James Watt's development of the first practical steam engine. Watt, whose name is now the *Systeme International des Unites* (S.I.) term for power, first used his engine to pump water out of mines. Previously, such pumping had been done with draft horses, so it was logical to relate the work the steam engine could do to the number of horses it could replace. The account presented here of how Watt did that is adapted from Herbert Arthur Kline's *The Science of Measurement: A Historical Survey*.

Watt's Draft Horse—The horse plodded a circular path, pulling at a right angle on the end of a 12-foot lever projecting from a capstan at the center of the circle. The capstan, in turn, was geared to operate the pump.

Watt estimated that the horse pulled with a force of 180 pounds. The circle it followed had a circumference of 2 times *pi* times a radius of 12 feet, or 75.398224 feet. The horse could make 144 trips around the circle in an hour or 2.4 trips a minute, for a speed of 180.95573 or about 181 feet per minute.

To convert that demonstration of the horse's ability into measurable leverage or what is commonly referred to as *torque*, Watt multiplied 180 pounds times 181 feet, obtaining 32,580 pounds-feet per minute. He rounded that figure up to 33,000 pounds-feet per minute, or 550 pounds-feet per second, which became the norm for 1.0 horsepower.

Generating Force—Watt's draft horse generated force around the circumference of a circle and, as the animal pulled on the lever, that force was applied to the capstan at the center of the circle. An automobile engine can be described as doing just the opposite. It delivers force at the output end of the crankshaft. Envision a 1.0-foot lever attached at a right angle to the crankshaft at that point. As the crank rotates, the free end of the lever will follow a circle with a radius of 1.0 foot.

Watt's definition of horsepower involved a force in pounds, applied over a distance in feet, for a time of 1.0 minute. Therefore, to convert the rotational force of the crankshaft into horsepower, you must know the distance the free end of the 1.0-foot lever will go in 1.0 minute. That, of course, would be the circumference of a circle with a 1.0-foot radius multiplied by the number of crankshaft revolutions per minute, or rpm.

The circumference is *pi* multiplied by 2 multiplied by 1.0 foot or, more simply, 2 times *pi*, which is 6.2831853 feet.

Therefore, the total distance the free end of the lever will go in 1.0 minute is 6.2831853 feet times rpm.

The product of that calculation can be multiplied by the known torque of the engine to find the total pounds-feet of torque per minute. The result can then be divided by Watt's pounds-feet figure per minute for 1.0 horsepower (33,000) to find the engine's horsepower. That works out to the following formula:

$$\text{Horsepower} = \frac{6.2831853 \times \text{rpm} \times \text{torque}}{33,000}$$

Dividing the right side of the equation by two times *pi*, you can eliminate 6.2831853 and reduce 33,000 to 5252.1131. By rounding down the latter figure, you can simplify the formula to:

$$\text{Horsepower} = \frac{\text{rpm} \times \text{torque}}{5252}$$

Here is the control panel of a classic Heenan and Froude G 240 EH dynamometer, a costly unit imported from England to the U.S. in the 1960s and 1970s. It was capable of handling up to 1000 horsepower. This particular dyno was installed at Ronnie Kaplan Engineering in Elk Grove, Illinois, in the late 1960s.

This is the formula used to determine horsepower when an engine is tested on a dynamometer. On modern computerized dynos, measurements and calculations are done electronically. On older units, though, torque was found by measuring the resistance of a device known as a *Prony brake* against the flywheel end of the crankshaft. Therefore, output figures at the flywheel are still called *brake* torque and *brake* horsepower, or *bhp*, after the old Prony brake.

GROSS VS. NET RATINGS

There are two forms of brake torque and brake horsepower, gross and net. The gross figures represent what the engine can do under ideal conditions, with "laboratory" intake and exhaust manifolding, and without the load of any auxiliary equipment except the fuel, oil and water pumps. They show the absolute maximum output at the flywheel.

The net torque and power, on the other hand, represent, what the engine can do as it's installed in a vehicle, with such auxiliary items as the air cleaner, alternator, fan and standard intake and exhaust systems in place.

Musclecar Ratings—During the 1950s and '60s, automakers tried to outdo each other in gross horsepower claims. By the late '60s, that policy had begun to backfire as insurance companies began adding expensive surcharges to premiums on cars with high-powered engines. So the factories reversed themselves and actually underrated their hottest powerplants.

They underrated the brake horsepower, that is, not the brake torque. A classic example was the street version of the Chrysler 426 cubic-inch Hemi. Its advertised output was 425 bhp at 5000 rpm, and the dyno showed 446 pounds-feet of torque at 5000 rpm. Applying the formula with those figures:

$$\text{Horsepower} = \frac{(5000 \times 446)}{5252}$$

The G 240 EH dyno was really a water pump which measured torque on the basis of the water pressure created by the engine. As output rose, the water could overheat and cavitate, making it impractical to measure more than the equivalent of 1000 horsepower.

In 1971, GM announced it was switching from gross to net horsepower and torque ratings in its advertised specifications. For that year, they published both. The differences were dramatic. For example, the 1971 442 W-30 went from a 350 gross horsepower rating to a 300 net horsepower rating! (courtesy Musclecar Review).

The result is 424.60015 which of course can be rounded up to 425, Chrysler's bhp figure. But Chrysler never claimed that was the maximum output! On the dyno, the rpm continued to climb faster than the torque dropped and, at 6000 rpm, the reading was 412 pounds-feet. Let's try the formula again:

$$\text{Horsepower} = \frac{(6000 \times 412)}{5252}$$

That works out to 470.6778 or, rounded up, 471 bhp! And that is 46 bhp, or almost 11 percent more than Chrysler officially claimed!

The practice of unrealistic engine output claims, whether exaggerated or underrated, began to fade in 1971, when General Motors announced it was switching from gross to net horsepower and torque ratings in its advertised specifications. Within the next two years, all domestic automakers and most importers followed suit, thus ending claims based on gross output.

For the 1971 model year, GM published both gross and net figures, the only time I'm aware of any manufacturer doing so. The differences were revealing. To take just one example, the base version of that year's 350 cubic-inch Chevrolet V-8 had a gross rating of 245 bhp, but a net rating of only 165 bhp—32 percent less!

Present-day enthusiasts sometimes ask if there is a conversion factor to determine the net equivalent of the gross horsepower ratings used before the early 1970s. Unfortunately, there isn't, because the gross ratings varied so widely—some of them way above the true gross output and some way below—and no single conversion factor could take all the variations into account.

Therefore, the only way to determine the original net output of a musclecar engine of the 1960s is to run that engine in its original form on a dyno.

CALCULATING TORQUE

If you know the horsepower at a given rpm, you can determine the torque at that same rpm by transposing the formula to:

$$\text{Torque} = \frac{(5252 \times \text{horsepower})}{\text{rpm}}$$

Using the advertised output of the Hemi:

$$\text{Torque} = \frac{(5252 \times 425)}{5000}$$

rpm provides a figure called the brake specific fuel consumption, *brake specific* for short or bsfc abbreviated. To restate that as a formula:

$$BSFC = \frac{\text{fuel pounds per hour}}{\text{brake horsepower}}$$

The lower the brake specific, the less fuel the engine is using to develop the horsepower in question; in other words, the more efficient the engine and the better the fuel economy.

Example—Suppose you have an engine which shows a fuel-flow rate of 144 pounds per hour at the rpm where it develops 300 brake horsepower:

$$BSFC = \frac{144}{300}$$

The brake specific would be 0.48. Generally, the figure should be 0.50 or less. It won't be constant. It will be at its lowest, i.e., most efficient, at or near the same rpm where peak torque is developed.

At lower rpm, the airflow through the intake system is slower than it is at peak torque and, at higher rpm, there isn't time to maintain the same airflow as there is at peak torque. So at engine speeds below or above peak torque, airflow will not be as efficient and, as a result, the brake specific fuel consumption will be greater.

DYNO CHART

When an engine is tested on a dynamometer, a chart is prepared of the brake torque in pounds-feet and the brake horsepower at specific rpm intervals. An example appears in Figure 4a on page 29. It shows the torque and horsepower for a modified 350 cubic-inch Chevy V-8 at intervals of 200 rpm from 3000 to 7000 rpm.

Maximum torque, 350 pounds-feet, occurs from 3800 to 4000 rpm and maximum horsepower, 343, at 6000 rpm. Those are all relatively high figures for a small V-8, indicating that the engine has been reworked for higher performance.

Note how flat the torque output is between 3000 and 4200 rpm. Over that 1200-rpm spread, it varies only 10 pounds-feet. Beyond 4000 rpm, the torque gradually declines, though it hits another flat spot of 305 pounds-feet from 5400 to 5800 rpm.

But, as rpm continues to climb, so does horsepower, until the engine reaches its peak output at 6000 rpm. From

Ronnie Kaplan (foreground) and associate Pat Pallasch did not get a direct horsepower reading from the G 240 EH. They had to calculate it from torque, using the formula described in the text.

The answer is with 446.42 pounds-feet of torque at 5000 rpm, which is essentially the same as the reading produced on the dyno.

BRAKE SPECIFIC FUEL CONSUMPTION

Another aspect of engine performance and efficiency measured on a dynamometer is *fuel consumption*. An engine on a dyno is stationary and its fuel consumption obviously can't be measured in miles per gallon. However, if the dyno is equipped with a fuel flow meter, it will show the rate of fuel flow in pounds per hour. The rate at any given rpm divided by the brake horsepower at the same

The exotic headers on this Chevy 454 typify the alterations to stock specs and equipment that were allowed when testing for gross horsepower. That practice waned in the early 1970s and today's engines are rated at net horsepower, which represents their output as actually installed in a car.

that point, torque starts to fall sharply, while horsepower drops more slowly.

Most dyno charts will show much more than the one in Figure 4a. They'll include not only torque and horsepower, but fuel flow and brake specific fuel consumption, along with intake and exhaust specs and ignition settings.

But the real guts of dyno testing will be seen in the relationships of rpm, torque and horsepower.

EFFECTS OF ELEVATION

When you drive into the mountains in a car with a normally aspirated engine—that is, one without forced induction such as turbocharging—you'll notice that the higher you go above sea level, the less power you'll have. As the air thins out, less of it will be drawn into the engine. There will be a decrease in volumetric efficiency (a topic I'll discuss further in Chapter 6) and, with it, a loss in engine output. The resulting drop in horsepower at any

given rpm is approximately 3.0 percent for each 1000 feet of elevation.

If you live in flatlands at or near sea level, none of this will be of much significance to you. But, if you have ever driven over California's Sierras or Colorado's Rockies, you know what I'm talking about.

Example—Let's consider a trip to Colorado in a late-model Chrysler K-car with a 2.5-liter, 4-cylinder engine developing 100 net bhp. That nice, even output figure will simplify the calculations.

From the East or Midwest, your first major stop in Colorado is likely to be the mile-high city of Denver. On the way there, you'll begin to feel the effects of elevation as you leave the plains of the Midwest and start climbing the Eastern slope of the Rockies. When you reach Denver, you'll be at over 5000 feet and, at a 3.0 percent loss per 1000 feet, that means you'll have lost 15 percent of your power. The 100-bhp engine now has only 85 bhp.

RPM	LB-FT	BHP
3000	340	194
3200	340	207
3400	345	223
3600	345	236
3800	350	253
4000	350	267
4200	340	272
4400	335	281
4600	330	289
4800	325	297
5000	315	300
5200	310	307
5400	305	314
5600	305	325
5800	305	337
6000	300	343
6200	280	331
6400	255	311
6600	240	302
6800	190	246
7000	160	213

But that's just for openers. Heading west from Denver on Interstate 70, you'll cross the Rockies over passes at elevations of 10,000 and even 11,000 feet. And, at 10,000 feet, the engine will lose 30 percent of its output, which will decrease to 70 bhp.

Or suppose you head South to Colorado Springs to try climbing Pikes Peak, one of the highest spots in the United States that can be reached by automobile. At the summit, you'll be more than 14,000 feet above sea level. That will cost you 42 percent of engine power, dropping the little 100-bhp four-banger's output to only 58 bhp!

Of course, if you'd opted for a turbo when you ordered the K car, you might've avoided such losses. But that's a whole 'nother story.

Figure 4a
This chart shows the torque and horsepower for a modified 350 cubic-inch Chevy at intervals of 200 rpm from 3000 to 7000 rpm as measured on a dyno.

Table 4

FORMULAS FOR BRAKE HORSEPOWER & TORQUE

$$\text{horsepower} = \frac{\text{rpm x torque}}{5252}$$

$$\text{torque} = \frac{5252 \text{ x horsepower}}{\text{rpm}}$$

$$\text{brake specific fuel consumption} = \frac{\text{fuel pounds per hour}}{\text{brake horsepower}}$$

$$\text{bhp loss} = \frac{\text{elevation in feet}}{1000} \text{ x } 0.03 \text{ x bhp at sea level}$$

Indicated Horsepower & Torque 5

Brake torque and horsepower represent output at the flywheel as measured on a dynamometer. However, those figures do not account for losses within the engine due to inertia and friction. Therefore, the brake figures will always be less than the horsepower and torque actually developed within the cylinders. Once you know the flywheel and cylinder outputs, you can determine the engine's true mechanical efficiency. An indicator is used to measure cylinder pressure during each of the four strokes—intake, compression, combustion and exhaust—and from them the indicated mean effective pressure, or mep, can be determined. Once you know the mep, indicated horsepower and torque can be calculated (photo by Michael Lutfy).

As demonstrated in the previous chapter, brake torque can be measured on a dynamometer and, from it, brake horsepower can be calculated. However, those figures represent output at the flywheel and, because of losses within the engine primarily from friction and also from inertia, they will always be less than the horsepower and torque actually developed within the cylinders. And, once the output both at the flywheel and in the combustion cylinders is known, you can determine the engine's mechanical efficiency.

31

INDICATED MEAN EFFECTIVE PRESSURE

You can't measure the horsepower and torque developed within the cylinders directly. However, using a device called an *indicator*, you can measure cylinder pressure during each of the four strokes—intake, compression, combustion and exhaust—and, from them, you can find the *indicated mean effective pressure*, or *mep*, which is a form of output that occurs within the cylinders and is unaffected by friction and inertia. When the indicated mep is known, it's possible to calculate the indicated horsepower and torque within the cylinders.

An indicator is not something the average hot rodder is likely to have readily available. Nonetheless, the serious performance enthusiast should be aware of the interrelationships of mep, horsepower and torque. (It should be noted that the formulas involving these interrelationships are equally valid for either brake or indicated figures.)

INDICATED HORSEPOWER

There's a formula for calculating horsepower from mep that's favored by many engineering theorists because it involves a simple acronym, **PLAN**, that's easy to remember:

$$\text{Horsepower} = \frac{P \times L \times A \times N}{33,000}$$

P stands for mep in pounds per square inch or psi; **L** for the length of the stroke in feet; **A** for the top surface area of one piston in square inches; and **N** for the number of power strokes per minute. When these four factors are multiplied together, they show the total amount of torque the engine develops in one minute. That figure is then divided by 33,000—the number of pounds-feet per minute equal to one horsepower—to find the total horsepower.

Part of the appeal of **PLAN** is that it focuses on the aspects of engine design that ultimately determine horsepower. When you modify an engine to improve performance, you are really increasing **P, L, A** and/or **N**.

Example—For example, when you raise the compression ratio, **P** is increased. When an engine is bored or stroked, **A** or **L** is increased. When you regrind the crankshaft to get higher rpm, you increase **N**.

The formula is awkward in practical use, though, because it requires separate calculations to find **L**, **A** and **N**. To find **L**, the stroke in inches is divided by 12 or:

$$L = \frac{\text{stroke}}{12}$$

To find **A**, divide *pi* by 4 then multiply it by the bore squared, or:

$$A = \frac{pi}{4} \times \text{bore}^2$$

And, to find **N** for a conventional four-stroke-cycle engine, divide the rpm by 2 and multiply the result by the number of cylinders, or:

$$N = \frac{\text{rpm}}{2} \times \text{no. of cylinders}$$

So our "simple" acronym eventually leads to a much more complicated formula:

$$\text{Horsepower} = \frac{\text{mep} \times \text{stroke} \times \text{bore}^2 \times pi \times \text{rpm} \times \text{no. of cylinders}}{12 \times 4 \times 2 \times 33,000}$$

But look closely. Embedded in there are the factors of the formula for displacement:

$$\text{Displacement} = pi/4 \times \text{bore}^2 \times \text{stroke} \times \text{no. of cylinders}$$

You can remove those and use the displacement in their stead, simplifying the raw formula to:

$$\text{Horsepower} = \frac{\text{mep} \times \text{displacement} \times \text{rpm}}{12 \times 2 \times 33,000}$$

And that, of course, can be reduced still further to:

$$\text{Horsepower} = \frac{\text{mep} \times \text{displacement} \times \text{rpm}}{792,000}$$

Example—Let's take as an example a 302 cubic-inch engine which has an indicated mep of 175 psi at 4200 rpm:

$$\text{Horsepower} = \frac{175 \times 302 \times 4200}{792,000}$$

Working the equation through gives an answer of 280.265 indicated horsepower.

INDICATED TORQUE

The mep will be in direct proportion to the torque, and the peak mep will occur at the same rpm as the peak torque. To find indicated torque, multiply the mep times the displacement, and divide the result by *pi* times 4 times 12 or:

$$\text{Torque} = \frac{\text{mep} \times \text{displacement}}{3.1415927 \times 4 \times 12}$$

Multiplied, the constants become 150.79645, which can be rounded off to 150.8, reducing the formula to:

$$\text{Torque} = \frac{\text{mep} \times \text{displacement}}{150.8}$$

Example—Assume that the 302 cubic-inch engine has a maximum indicated mep of 177.5 psi at 3200 rpm—though the engine speed isn't immediately relevant:

$$\text{Torque} = \frac{177.5 \times 302}{150.8}$$

And that would mean a maximum indicated torque of 355.5 pounds-feet.

BRAKE MEAN EFFECTIVE PRESSURE

Mean effective pressure occurs within an engine's cylinders and cannot be directly measured at the flywheel, as brake torque is. However, you can find mep from either horsepower or torque. If you start with brake horsepower or brake torque, you will come up with a hypothetical brake mean effective pressure, or bmep.

To find bmep from horsepower, the formula is:

$$\text{MEP} = \frac{\text{hp} \times 792,000}{\text{displacement} \times \text{rpm}}$$

Example—Let's see what the bmep would be at the 7000-rpm horsepower peak of that 426 cubic-inch, 471-bhp Chrysler Hemi I described on page 25 in Chapter 4:

$$\text{MEP} = \frac{471 \times 792,000}{426 \times 7000}$$

The answer is 125 psi. The formula for finding bmep from torque is:

$$\text{MEP} = \frac{\text{torque} \times 150.8}{\text{displacement}}$$

Chrysler listed the 426 Hemi's maximum brake torque as 490 pounds-feet at 4000 rpm. Therefore, the input for bmep would be:

$$\text{MEP} = \frac{490 \times 150.8}{426}$$

That results in a figure of 173 psi.

BMEP vs. BSFC—Note that the bmep is higher at peak torque than at peak horsepower. Like brake specific fuel consumption, discussed in Chapter 4, brake mean effective pressure is at its best at peak torque. Indeed, as measures of an engine's efficiency, the bsfc and bmep tend to reflect each other. The bsfc is at its lowest or most efficient at peak torque, and the bmep is at its highest or most efficient at the same point. On either side of peak torque, the bsfc gets worse as it increases and the bmep gets worse as it decreases.

While bmep is a calculated, theoretical form of output, it is useful for comparing the relative performances of different engines. It is entirely possible, for example, that a small 4-cylinder sports car engine and a big stock car racing V-8 could have similar bmep figures, despite vast differences in their displacements and their horsepower and torque characteristics.

Typical bmep figures are 130 to 145 psi for the engine in a standard passenger car, 165 to 185 psi in a high-performance or sports car, and 185 to 210 psi in a racing vehicle.

MECHANICAL EFFICIENCY

An important advantage of having both indicated and brake output figures is that they can be used to determine the engine's percentage of mechanical efficiency.

The basic formula is the same, whether the figures used are horsepower, torque or even mep:

$$\text{Mechanical Efficiency} = \frac{\text{brake output}}{\text{indicated output}} \times 100$$

Mechanical Efficiency from Horsepower

—Let's take that 302 cubic-inch engine which was found to have 280.265 indicated horsepower at 4200 rpm and 355.5 pounds-feet of torque at 3200 rpm, and assume that its brake output ratings are 225 horsepower and 300 pounds-feet at the same respective engine speeds. To find mechanical efficiency from the horsepower figures, you use:

$$\text{Mechanical Efficiency} = \frac{225}{280.265} \times 100$$

And the result would be 80.28 percent mechanical efficiency. The difference between the two horsepower figures—indicated output minus brake output or, in this case, 55.265—is known as *friction horsepower*, because it is the amount lost between the cylinders and the flywheel from friction.

Mechanical Efficiency from Torque

—To find mechanical efficiency from the brake and indicated torque ratings of our 302-cubic-inch engine, the figures would be:

$$\text{Mechanical Efficiency} = \frac{300}{355.5} \times 100$$

This time, you have 84.39 percent mechanical efficiency and 55.5 pounds-feet of friction torque. And, once again, here's an example of greater efficiency at peak torque than at peak horsepower.

TAXABLE HORSEPOWER

There's a final, theoretical form of horsepower you should be aware of, even though it hasn't been taken seriously for nearly 50 years. That's the *rated* or *taxable horsepower*, using a formula developed in the early 20th century and promulgated by the Society of Automotive Engineers (SAE) in this country and by the Royal Automobile Club in England.

The formula called for multiplying the square of the bore by the number of cylinders and dividing the result by 2.5, or:

$$\text{Taxable Horsepower} = \frac{\text{bore}^2 \times \text{cylinders}}{2.5}$$

If your 302 cubic-inch V-8 had a bore of 4.0 inches, its taxable output would be:

$$\text{Taxable Horsepower} = \frac{4^2 \times 8}{2.5}$$

That works out to 51.2 horsepower, a far cry from the indicated or brake figures.

The formula for taxable horsepower didn't directly consider piston stroke or rpm, though it assumed an mep of 90 psi and piston speed of 1000 feet per minute, reasonable figures in the fledgling years of the automobile but well below the capabilities of modern engines.

Nonetheless, as recently as the 1940s, taxable horsepower was used in some of the United States and in England as the basis not only for taxes on cars but for insurance rates as well. It also served as a model designation for several British cars, such as the famous Austin Seven which had a small-bore, 4-cylinder engine with approximately 7.0 taxable horsepower.

But it retarded modern engine development, particularly in England, by encouraging small-bore, long-stroke engines. By holding down the size of the bore (or A in the acronym PLAN), the engine designer hoped the engine would qualify for a low tax rate. He would compensate for the small bore by using a longer stroke (L) in hopes of increasing the mep (or P) for higher indicated and brake outputs. But that longer stroke, in turn, meant higher piston speeds and, with them, greater engine wear.

All this became academic by the 1940s as the use of taxable horsepower for taxation and insurance purposes began to disappear, clearing the way for modern big-bore, short stroke powerplants.

Table 5

FORMULAS FOR INDICATED HORSEPOWER & TORQUE

$$\text{horsepower} = \frac{\text{mep x displacement x rpm}}{792{,}000}$$

$$\text{torque} = \frac{\text{mep x displacement}}{150.8}$$

$$\text{mep} = \frac{\text{hp x 792,000}}{\text{displacement x rpm}}$$

$$\text{mep} = \frac{\text{torque x 150.8}}{\text{displacement}}$$

$$\text{mechanical efficiency} = \frac{\text{brake output}}{\text{indicated output}} \times 100$$

$$\text{friction output} = \text{indicated output - brake output}$$

$$\text{taxable horsepower} = \frac{\text{bore}^2 \text{ x cylinders}}{2.5}$$

Air Capacity & Volumetric Efficiency

6

To determine what size carburetor you'll need for your street engine, carburetion authorities recommend you assume a volumetric efficiency of 85% and use that in the formula for theoretical air capacity to find the proper carburetor flow in cfm (photo by Michael Lutfy).

An automobile engine is a form of air pump, and knowing its theoretical air capacity is necessary to determine its *volumetric efficiency*, i.e., the relationship between the theoretical capacity and the actual airflow. In addition, on a carbureted engine, the air capacity may serve as a guide to choosing the proper carburetor size.

AIR CAPACITY

The air capacity is a product of rpm and displacement. In a conventional four-stroke engine, the volume displaced on intake strokes during each crankshaft revolution will be 1/2 of the overall cubic capacity. So, to find the air capacity in cubic inches per minute, multiply the rpm by the displacement in cubic inches and divide by 2, or:

$$\text{Air Capacity} = \frac{\text{rpm x displacement}}{2}$$

In practice, calculating the air capacity in cubic inches per minute would result in unwieldy figures, so the measurement is converted to cubic feet per minute or cfm by dividing the displacement by 1728, the number of cubic inches in a cubic foot:

$$\text{CFM} = \frac{\text{rpm x displacement}}{2 \times 1728}$$

That, in turn, can be simplified to:

$$\text{CFM} = \frac{\text{rpm x displacement}}{3456}$$

Example—As an example, consider the theoretical air capacity of the 350 cubic-inch Chevy V-8 used for the dyno chart in Figure 6a on this page, the same chart used in Chapter 4. You are interested in two particular engine speeds: the rpm at peak torque because, like other measurements of engine efficiency already discussed, volumetric efficiency is highest at that point; and the maximum rpm, because that's where the air capacity is its greatest.

According to the chart, maximum torque is delivered at 4000 rpm, so to find the air capacity:

$$\text{CFM} = \frac{4000 \times 350}{3456}$$

Which works out to 405 cfm. Maximum rpm on the chart is listed as 7000 rpm, so to determine what this engine's greatest air capacity is:

$$\text{CFM} = \frac{7000 \times 350}{3456}$$

The answer is 709 cfm.

RPM	LB-FT	BHP
3000	340	194
3200	340	207
3400	345	223
3600	345	236
3800	350	253
4000	350	267
4200	340	272
4400	335	281
4600	330	289
4800	325	297
5000	315	300
5200	310	307
5400	305	314
5600	305	325
5800	305	337
6000	300	343
6200	280	331
6400	255	311
6600	240	302
6800	190	246
7000	160	213

Figure 6a. This chart shows the torque and horsepower for a modified 350 cubic-inch Chevy at intervals of 200 rpm from 3000 to 7000 rpm as measured on a dyno.

VOLUMETRIC EFFICIENCY

The actual airflow can be measured at each rpm and then divided by the theoretical capacity at the same rpm to find the engine's volumetric efficiency, or V.E. The resulting figure can be multiplied by 100 to convert it from a decimal to percent. Stated as an equation, that becomes:

$$\text{V.E.} = \frac{\text{actual cfm}}{\text{theoretical cfm}} \times 100$$

Let's suppose the actual airflow of our engine was 365 cfm at 4000 rpm and 565 cfm at 7000 rpm. The volumetric efficiency at 4000 would be:

$$\text{V.E.} = \frac{365}{405} \times 100$$

Or just over 90 percent. And, at 7000 rpm with an actual airflow of 565 cfm:

$$V.E. = \frac{565}{709} \times 100$$

The volumetric efficiency at that rpm is just under 80 percent.

According to Mike Urich, former engineering vice president of Holley Carburetors and author of HPBooks' *Holley Carburetors and Manifolds*:

"An ordinary low-performance engine has a V.E. of about 75 percent at maximum speed; about 80 percent at maximum torque. A high-performance engine has a V.E. of about 80 percent at maximum speed; about 85 percent at maximum torque. An all-out racing engine has a V.E. of about 90 percent at maximum speed; about 95 percent at maximum torque. A highly tuned intake and exhaust system with efficient cylinder-head porting and a camshaft ground to take full advantage of the engine's other equipment can provide such complete cylinder filling that a V.E. of 100 percent—or slightly higher—is obtained at the speed for which the system is tuned."

An engine would have to have a great air capacity to run this Holley 1050 cfm Dominator carburetor (courtesy Holley Carburetors).

Holley designed the 4010 four-barrel carburetor for performance. It's available in 600 cfm and 750 cfm versions (courtesy Holley Carburetors).

In practice, a highly tuned engine of the type Urich mentions may have a V.E. as much as 10 or 12 percent greater than the theoretical air capacity.

CARBURETOR SIZE

In an era when most high-performance engines are being equipped with fuel injection, carburetion may seem old fashioned. But it's not dead yet, and it's likely to be around for some time to come, although maybe not on new cars. There are still a lot of hot rods being built with carburetors. There are older cars being restored, such as musclecars of the 1960s, with carburetors. And in several forms of racing, there are classes that continue to require carburetors rather than injectors. So, to a lot of performance enthusiasts, carburetion remains important.

An engine's air capacity can be a guide to choosing carburetor size, in terms of the carburetor's airflow in cfm, for a given combination of displacement and rpm. This time, though, we're concerned with practical rather than theoretical capacity, i.e., the actual airflow.

Street Carb—But how do you estimate what size carburetor you need while you're building an engine, before you can measure the airflow? For a street engine, Urich and most other authorities on carburetion recommend assuming a volumetric efficiency of 85 percent and plugging that figure into the formula for theoretical air capacity to find the proper carburetor flow in cfm:

This Dominator by Holley has a capacity of 750 cfm. According to the example in the text for calculating racing carburetor cfm, this carburetor would be a good choice (courtesy Holley Carburetors).

$$\text{Street Carb cfm} = \frac{\text{rpm} \times \text{displacement}}{3456} \times 0.85$$

Example—Suppose you are building a street rod with a Chevy 350 V-8 that you don't expect to rev beyond 5000 rpm.

$$\text{Street Carb cfm} = \frac{5000 \times 350}{3456} \times 0.85$$

Before taking the V.E. into account, that would work out to 506 cfm. Multiply by that 0.85 and the figure drops to only 430 cfm.

Racing Carb—For a street engine, that formula is fairly reliable. But on a racing engine, as Urich points out, the volumetric efficiency can increase to 100 percent or more. Let's take the case of a highly tuned engine which can compress the air somewhat and increase the actual flow to 10 percent more than the theoretical air capacity. That's the same as saying 110 percent of, or 1.1 times, the theoretical figure:

$$\text{Racing Carb cfm} = \frac{\text{rpm} \times \text{displacement}}{3456} \times 1.1$$

For this example, suppose you have a Chevy 350 built for racing that you expect to be running up to 7000 rpm:

Mike Urich, a former engineer at Holley carburetors and author of HPBooks' Holley Carburetors and Manifolds, tested a Camaro with a 350 cubic-inch engine tuned for drag racing. Passes were made with a 650 cfm and 800 cfm carburetor. The car was quicker both in top speed and elapsed time with the 650 cfm carburetor, proving that bigger isn't always better (photo by Bob McClurg).

$$\text{Racing Carb cfm} = \frac{7000 \times 350}{3456} \times 1.1$$

This time, the initial figure is 709 cfm which, multiplied by 1.1, becomes 780 cfm.

However, if you go very far beyond the theoretical air capacity in choosing carburetor size, you'll quickly reach a point of diminishing returns. Mike Urich cites an instructive illustration in a drag racing test of a Chevrolet Camaro with the 350-cubic-inch engine. Using a 650-cfm Holley carburetor, the car ran the 1/4-mile in an elapsed time of 14.35 seconds with a terminal speed of 96.05 mph. Switching to a bigger 800-cfm Holley, the ET increased to 14.61 seconds and the speed dropped to 95.44 mph. Those may not seem like significant differences, but drag races are often won or lost by even narrower margins.

BIGGER IS NOT ALWAYS BETTER

Carburetors are not made in an infinite variety of sizes, so you're not likely to find one that corresponds exactly with your calculated carburetor cfm. In most cases, you'd probably be wiser to choose a carburetor the first available size down from your calculated figure rather than the first size up.

As a case in point, suppose you determine that you need a 720-to-730-cfm carburetor. However, the make and model of carb you want to use is available in either 700- or 750-cfm sizes, with nothing in between. Your temptation might be to go for the 750, but you'd probably be better off with the 700.

For choosing a carburetor is one area where bigger is not always better.

Table 6

FORMULAS FOR AIR CAPACITY & VOLUMETRIC EFFICIENCY

$$\text{theoretical cfm} = \frac{\text{rpm x displacement}}{3456}$$

$$\text{volumetric efficiency} = \frac{\text{actual cfm}}{\text{theoretical cfm}} \times 100$$

$$\text{street carb cfm} = \frac{\text{rpm x displacement}}{3456} \times 0.85$$

$$\text{racing carb cfm} = \frac{\text{rpm x displacement}}{3456} \times 1.1$$

It's possible to use a commercial truck scale, or a similar scale at a sand and gravel yard or a moving company. The problem is that the vehicle will have to be jockeyed around to get front/rear and left/right weight distribution figures, and that can be time-consuming.

Weight distribution is a statement of the percentages of a vehicle's overall weight divided lengthways between the front and rear wheels, or sideways between the left and right wheels. It is an important factor in the handling of all types of vehicles, from family sedans and race cars to motorhomes and heavy trucks.

The lengthways or front/rear weight distribution will vary greatly, depending on the type of vehicle. However, with the important exception of cars built for either oval track or drag racing, there's usually little if any significant difference in the sideways or left/right weight distribution.

WEIGHING THE VEHICLE

To find how the weight of a car or truck is distributed, you must first find the weights at the wheels. The ideal method is to weigh the vehicle with each wheel on a separate flat scale. A professional race team will often

The best way to weigh a vehicle is with an individual flatbed scale at each wheel. The measurements should be taken on level ground, with all four wheels on scales at the same time.

carry a set of four flat scales to check weights on the wheels when setting up a car for specific track conditions, but a set of such scales would be an extravagance for anyone not needing them regularly.

Public Scale—A public scale, such as one of those used by professional truckers or moving companies, is a less costly alternative, though it may also be less accurate than individual flat scales. The public scale has a capacity of several tons to weigh heavy trucks and their cargos, and at the more modest weights of ordinary cars or light trucks,

the device's readings may not be precise. For that reason, it isn't really worth the trouble to use such a scale to find the weight at each wheel.

Wheel Weights in Pairs—Fortunately, it isn't really necessary, either. The formulas involving weights at the wheels that I'll be discussing in this chapter and the next call for those weights in pairs—the fronts and rears and, for oval track cars, the lefts and rights—so those are the figures you should measure on the public scale.

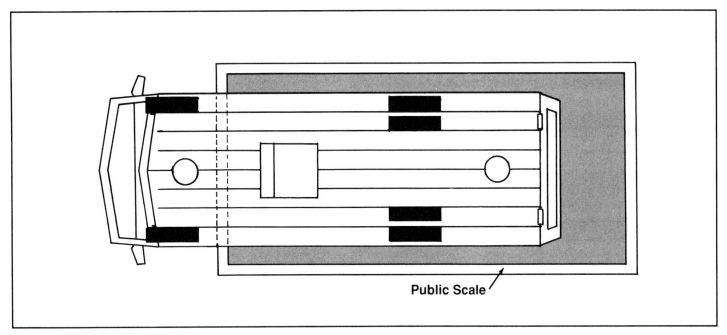

On a public scale, the weight on the wheels at one end of a vehicle can be found by moving the wheels at the opposite end just off the scale platform.

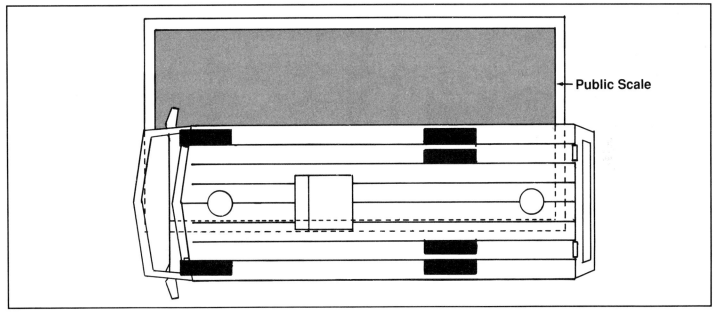

Similarly, the weight on the wheels at either side can be found by positioning the wheels at the other side just off the platform.

Measuring Weight—Start by finding the overall weight of the vehicle with all four wheels on the scale. Next, measure the weight with the wheels at one end of the vehicle on the scale and then the wheels at the other end. Finally, if you need to know the sideways weight distribution, weigh the wheels at one side of the vehicle and then the other side.

If the ramps at the ends or sides of the scale platform aren't level, position the vehicle with the wheels being weighed as far on to the platform as possible, and the wheels not being weighed as close to the edge without touching it as possible. If the vehicle is tilted at the slightest angle, there will be enough weight transfer to the wheels off the scale to invalidate the readings.

Isn't it redundant to weigh both ends? Couldn't you weigh the front wheels and then subtract that figure from the overall weight to find the weight at the rear wheels? Yes, you could. But by actually weighing both sets of wheels, you'll have an optimum check on the scale's accuracy because the overall weight and the sum of the front and rear weights should be the same. If you weigh the left and right wheels, their sum should be the same, too. If there are any significant discrepancies, go to another scale and start over.

Obviously, all this is going to be time-consuming and you don't want to be shuttling a race car back and forth on the scale platform with a line of impatient truckers waiting to check their loads before heading out on the highway. So it's a good idea to check ahead of time with the scale operator to find when the facility isn't likely to be busy.

FRONT/REAR DISTRIBUTION

To find the percentage of weight on a given set of wheels, divide the weight on those wheels by the overall vehicle weight, and multiply the result by 100, or:

$$\text{Wheel Weight Percentage} = \frac{\text{weight on wheels}}{\text{overall weight}} \times 100$$

Example—Let's suppose you have a car that weighs 4000 pounds overall with 2240 pounds on the front wheels:

$$\text{Wheel Weight Percentage} = \frac{2240}{4000} \times 100$$

The answer is 56 percent, which means that over one-half of the overall vehicle weight is locate over the front wheels. If the scale were accurate, it would have shown 1760 pounds on the rear wheels:

$$\text{Wheel Weight Percentage} = \frac{1760}{4000} \times 100$$

Or 44 percent. Of course, you could also have arrived at that figure by subtracting 56, the percentage on the front wheels, from 100.

Typical Front/Rear Weights—That front/rear weight ratio of 56/44 is typical for a full-size sedan with a front engine and rear drive, like a Chevrolet Caprice or Ford Crown Victoria. The engine is the heaviest single

On a full-size passenger car with a front-engine/rear-drive layout, such as the Chevrolet Caprice, the weight distribution will be about 56 percent front and 44 percent rear.

component in a motor vehicle and its location—front, rear or in between—strongly affects the weight distribution.

On a front-engined, rear-drive sports or GT car with only two seats, like the Mazda Miata, Chevrolet Corvette or Mercedes-Benz SL, the smaller passenger compartment allows the engine to be set back somewhat for a better balanced front/rear weight ratio of 51/49 or 52/48.

For a compact or mid-size car with a front engine and front drive, as is the case with most of today's popular cars, the front transaxle increases the weight on the front wheels, and the front/rear weight ratio is likely to be between 60/40 and 65/35.

Conversely, on a true rear-engined car, i.e., one with the engine behind the rear transaxle, like the Porsche 911 and its many variants, the front/rear weight ratio will be on the order of 40/60.

On a mid-engined car, i.e., one with the engine behind the passenger compartment but ahead of the rear transaxle, such as the Toyota MR2 or Acura NSX, the proportion is slightly less severe, but only slightly, at about 42/58.

Understeer & Oversteer—It's widely believed that a vehicle's weight distribution affects its handling characteristics. A front-heavy car is supposed to understeer, i.e., its front wheels will drift toward the outside of a turn.

A rear-heavy car, on the other hand, will reputedly oversteer, i.e., its rear wheels will drift to the outside of a turn.

That's an oversimplification, though. A vehicle's tendency to understeer or oversteer is the product of the tire slip angles. Those, in turn, are affected not only by weight distribution, but also by such factors as suspension design and tire size. For example, rear-heavy cars like the 911, MR2 and NSX, often have bigger tires at the rear than at the front to help neutralize any tendency to oversteer and, in normal driving, they may even understeer somewhat.

However, when any car is pushed beyond its limit of adhesion, centrifugal force is likely to take over and cause the tires at the heavier end to lose traction and skid or even spin toward the outside of a turn.

LEFT/RIGHT DISTRIBUTION

On an oval track race car with rear drive, there should be a weight bias toward the rear and toward the left side— toward the rear to increase weight transfer to the rear for better traction when accelerating and to minimize weight transfer to the front when braking, and toward the left to decrease weight transfer to the right as the car follows a continuous series of turns to the left.

On a front-engined, rear-drive two-seater like the Mazda Miata, the engine can be set back somewhat to provide a nearly even distribution of 51 percent front/49 percent rear.

The Toyota MR2 is a typical mid-engined car, with the powerplant between the passenger compartment and the rear wheels. Its weight distribution is approximately 42 percent front and 58 percent rear.

The Acura NSX is also mid-engined, and has slightly wider tires at the rear than at the front to offset the bias of 58 percent of its weight to the rear.

In road racing, a rearward weight bias is favored for better traction and thrust. However, the side-to-side weight distribution should be even for consistent cornering in either direction (photo by Michael Lutfy).

During the 1970s, Chrysler Corporation produced a kit for building a Dodge Aspen- or Plymouth Volare-based stock car for oval track racing. According to the factory parts catalog for this vehicle, 53 to 55 percent of the weight should be on the left side and 52 to 54 percent on the rear wheels.

Drag race cars, too, are often built with their weight toward the rear and the left—toward the rear, again, to increase weight transfer for better traction, and toward the left to counteract torque reaction in that direction.

Torque reaction occurs when the driver releases the clutch pedal and punches the accelerator; the pinion (the gear at the end of the driveshaft) will attempt to climb the ring (the gear at the axle), causing the left side of the vehicle to lift slightly. Added weight on the left rear wheel will help minimize the effect.

To find the left/right weight ratio, use the same formula you did for the front/rear ratio, only this time start with the weight on the left wheels rather than on the front ones.

Example—Let's consider a dirt track stock car weighing 3000 pounds overall, with 1410 pounds on the front wheels, 1590 pounds on the rear, 1590 pounds on the left, and 1410 pounds on the right. In this instance, the weight distribution lengthways and sideways would be the same:

$$\text{Wheel Weight Percentage} = \frac{1590}{3000} \times 100$$

Which works out to 53 percent at the rear and left, within the parameters Chrysler recommended, and front/rear and left/right weight ratios both of 53/47.

ADDING WEIGHT AT EITHER END

If you add weight at either end of a vehicle, it will obviously affect the weight distribution by increasing the percentage on the nearer pair of wheels and decreasing the percentage on the farther pair.

Examples of this would be adding a winch at the front of a four-wheel-drive truck, or a trailer hitch at the rear of a tow vehicle, or ballast to a race car.

For oval track racing, where the cars turn left only, there should be a weight bias toward the left as well as toward the rear.

With the added object in place, you can take the vehicle back to the scale. Or you can calculate the added weight on the nearer wheels. To do that, you must know the vehicle's wheelbase, the amount of added weight and the horizontal distance of the added weight from the centers of the nearer wheels. Specifically, you need to know the distance of the added weight's center of gravity from the centers of the wheels.

CG of Add-Ons—The center of gravity or *cg* of an object is the point around which the weight of the object is evenly balanced in every direction. In the case of a symmetrical or regularly shaped object like a winch, you can assume the cg is at the center of the object. However, with a complex or irregularly-shaped object like a trailer hitch, estimating the cg is not so easy. One suggestion: Before the object is installed, try to balance it on a sawhorse.

Once you have an estimate of the distance of the added object's cg from the centers of the wheels, you can find how much weight it's going to add to the nearer wheels with the formula:

$$\text{Wheel Weight Increase} = \left(\frac{\text{from wheels}}{\text{wheelbase}} \times \text{weight} \right) + \text{weight}$$

Adding a Winch at Front—Suppose you have a sport-utility vehicle which has a 107-inch wheelbase and, in standard form, weighs 4000 pounds, with 55 percent or 2200 pounds on the front wheels and 45 percent or 1800 pounds on the rear wheels.

If you install a winch at the front which weighs 110 pounds, and you estimate that the cg of the winch is 45 inches ahead of the front wheel centers, the figures in the formula will be:

$$\text{Wheel Weight Increase} = \left(\frac{45}{107} \times 110 \right) + 110$$

That works out to 156 pounds added at the front wheels.

For trailering, proper understanding of the weights of both the tow vehicle and trailer is important. The vehicle manufacturer will usually specify the maximum trailer weight the vehicle can handle. The tongue weight of the trailer, i.e. the weight at the hitch, should be between 10 and 15 percent of the trailer's overall weight.

The 110-pound weight of the winch has acted as a force on a lever, with the front wheel centers as the fulcrum, lifting a weight of 46 pounds from the rear wheels and transferring it to the front ones. The overall vehicle weight is now 4410 pounds, with 2356 pounds on the front wheels and 1754 pounds on the rear ones, and the weight distribution has become 57/43.

TRAILER TONGUE WEIGHT

One of the most useful applications of the formula for added weight at the front or rear of a vehicle is to find the effect of the tongue weight of a trailer on the weight distribution of a tow car or truck.

I'm talking here about a vehicle with a *weight-carrying hitch*, which supports the entire tongue weight on the hitch ball, as opposed to a *weight-distributing hitch*, which uses springs and levers to divide the tongue weight between the front and rear wheels of the tow vehicle; the weight will lower the vehicle's ride height slightly, but the vehicle will remain level even with a heavy trailer attached. A weight-carrying hitch is satisfactory with smaller, lighter trailers, but a weight-distributing one is recommended for heavier towing.

The tongue weight of a trailer should be between 10 and 15 percent of the overall weight. For example, a 1000-pound trailer should have a tongue weight between 100 and 150 pounds.

Using a Bathroom Scale—How can you find the trailer's tongue weight? If the trailer has a gross weight rating of 2000 pounds or less, the tongue weight should be no more than 300 pounds and can be checked on an ordinary bathroom scale. However, if there's any reason to believe that the tongue weight is going to be beyond the 300-pound capacity of the typical bathroom scale, you'll have to use a little trickery.

Take a brick or a block of wood the same height as the scale and a 2x4 more than 3 feet long. Place the block so that its center line is exactly 1 foot to one side of the center line of the jack extension on the trailer tongue. Place the scale so that its center line is 2 feet to the other side of the

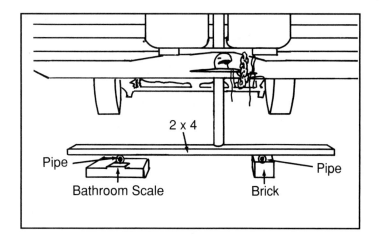

An ordinary bathroom scale with a capacity of 300 pounds can be used to check trailer tongue weights as high as 900 pounds by offsetting the scale from the point at which the weight is applied. The exact procedure is described in the text.

center line of the jack extension. Now place the 2x4 so that it stretches across both the block and scale. Lower the jack until the trailer tongue drops far enough that its weight is supported by the 2x4.

Read the scale and multiply the figure it shows by the number of feet, in this case 3, between the centers of the block and scale. That will give you the tongue weight.

The setup I've described will handle tongue weights up to 3 times the scale's capacity, or 900 pounds with a 300-pound scale. For heavier tongue weights than that, you can use a longer 2x4, move the scale out another 1 or 2 feet, and change the multiplication factor accordingly. The factor must be the same as the distance in feet between the centers of the block and scale.

Heavier Trailers—You can also take the trailer to a truck scale. However, because of such a scale's questionable accuracy with lower weights, you may not get satisfactory results by simply dropping the trailer tongue on the scale. In the words of Bill Estes, editor and associate publisher of *Trailer Life Magazine*:

"You'll get a much better figure by subtracting the difference, say, between 4000 and 4500 pounds than by weighing a 500-pound trailer tongue on a scale that may be calibrated for as much as 80,000 pounds."

Estes describes the procedure for finding the tongue weight:

"Position the car and trailer so the tongue jack and trailer wheels are on the scale but the car wheels are off. . . . Get a weight figure on the trailer wheels with the tongue resting on the tow vehicle off the scale. Then, lower the tongue jack to the scale, raise the coupler off the ball, drive the vehicle away and leave the trailer on the scale. You'll have two readings. Subtract to determine hitch weight."

Subtract, that is, the weight of the trailer hitched from the weight unhitched.

If the tow vehicle has a weight-distributing hitch, the spring levers should be disconnected to deactivate the weight-distributing feature and the back of the tow vehicle should be blocked or jacked so that the vehicle and trailer remain at their normal ride height.

Okay, suppose you have a 3000-pound trailer and you find it has a tongue weight of 450 pounds. For a tow vehicle, let's go back to that 107-inch-wheelbase sport-utility and assume that its 4000-pound weight includes a weight-carrying hitch. The center line of the hitch ball is 52 inches aft of the center lines of the rear wheels. Plugging the appropriate figures into the formula:

$$\text{Wheel Weight Increase} = \left(\frac{52}{107} \times 450 \right) + 450$$

You learn that 669 pounds have been added to the rear wheels, of which 669 minus 450, or 219 pounds, have been lifted from the front wheels. The front-wheel weight has been decreased from 2200 to 1981 pounds and the rear-wheel weight has gone up from 1800 to 2469 pounds. Overall weight is now 4450 pounds. Most importantly, what's happened to the weight distribution? Let's check at the front wheels:

$$\text{Wheel Weight Percentage} = \frac{1981}{4450} \times 100$$

That's 44.5 percent, which gives a front/rear weight ratio of 44.5/55.5, almost the reverse of the original 55/45!

That drastic a shift in weight distribution will cause the rear end of the tow vehicle to drop and the front end to rise, and the vehicle will ride at a noticeable angle instead of levelly. The added weight at the rear is likely to result in greater rear end sway around turns, making the vehicle more difficult to control. In addition, the headlights will point slightly upward, right into the eyes of drivers of oncoming vehicles.

And that helps explain why, with a heavier trailer, a weight-distributing hitch is preferable to a simple weight-carrying one.

Table 7

FORMULAS FOR WEIGHT DISTRIBUTION

$$\text{percent of weight on wheels} = \frac{\text{weight on wheels}}{\text{overall weight}} \times 100$$

$$\text{increased weight on wheels} = \left(\frac{\text{distance of cg from wheels}}{\text{wheelbase}} \times \text{weight} \right) + \text{weight}$$

Center of Gravity 8

In a race car, it's important to know the position of the center of gravity—longitudinally, laterally and vertically—in order to determine such aspects of vehicle behavior on the track as weight transfer to the rear when accelerating or to the outside of a turn when cornering (photo by Michael Lutfy).

The center of gravity or *cg* of an object, as noted in Chapter 7, is the point around which the weight of the object is evenly balanced in every direction. In the previous chapter, the concern was with the cgs of add-on devices like winches and hitches and the effects they have on weight distribution. In this chapter, I'm going to discuss the cg of the vehicle itself. This is especially important on a race car, where you need to know the exact location of the cg in order to predict several aspects of vehicle dynamics, such as how much weight transfer occurs to the rear while accelerating or to the outside of a turn while cornering.

HORIZONTAL POSITION

The horizontal position of the cg, both lengthways and sideways, is inversely proportional to the weight distribution. The horizontal position is measured in relation to the distances between the points at which the vehicle is weighed, i.e., the wheels.

Lengthways Location

Lengthways Location—The lengthways location of the center of gravity is measured as a part of the *wheelbase*. To find how far it is behind the front wheel centers, divide the weight on the rear wheels by the overall vehicle weight and then multiply the resulting decimal figure by the wheelbase, or:

$$\text{cg location behind front wheels} = \frac{\text{rear wheel weight}}{\text{overall weight}} \times \text{wheelbase}$$

As an example, go back to the oval track race car described in Chapter 7. That car weighs 3000 pounds overall, with 1410 pounds on the front and right wheels and 1590 pounds on the rear and left wheels. Now suppose that the wheelbase is 108 inches and the track 63 inches.

The information needed to find the lengthways position of the cg is shown in Fig. 8a on page 62 and is applied in the formula thusly:

$$\text{cg location behind front wheels} = \frac{1590}{3000} \times 108$$

cg location behind front wheels = 0.53 x 108

The cg is 57.24 inches behind the front wheel centers, as shown in Fig. 8b on page 62.

Sideways Location—Just as the lengthways position of the cg can be measured as a part of the wheelbase, the sideways location can be measured as a part of the vehicle *track*, which is the lateral distance between the centers of the treads of tires on either side. However, the sideways location of the cg is usually described in terms of how far it

Professional race teams often have sets of four scales, one for each wheel, to weigh their cars. Shown here is the setup used by Nissan Performance Technology (NPTI) to weigh its championship-winning Nissan GTP ZX (photo by Michael Lutfy).

58

is off-center toward the heavier side. To find that, divide the weight on the lighter side by the overall weight and multiply the resulting decimal by the track, then subtract that figure from 1/2 the track, or:

$$\text{cg location off-center on heavy side} = \frac{\text{track}}{2} - \left(\frac{\text{weight on light side}}{\text{overall weight}} \times \text{track} \right)$$

If there's a significant difference between the front and rear tracks, use the average of the two.

The sideways location in our oval track car would be:

$$\text{cg location off-center on heavy side} = \frac{63}{2} - \left(\frac{1410}{3000} \times 63 \right)$$

cg location off-center on heavy side = 31.5 − (0.47 × 63)

cg location off-center on heavy side = 1.89 inches

In this case, the cg is located 1.89 inches off-center to the left.

VERTICAL POSITION

On an automobile with a front-mounted, pushrod V-8 engine and rear-wheel drive, which is the configuration of most traditional high-performance cars, the cg will usually be from 14 to 22 inches above the ground. On such a car, one rule of thumb is that the cg will be at about the same height as the camshaft. To find what that height is, use a yardstick to measure from the ground to the camshaft centerline at the front of the engine.

However, to pinpoint the cg height more precisely, or to find it at all on a vehicle with other than a front V-8 engine/rear-drive layout, you'll have to do some jacking around—literally.

Measuring the weight at each wheel makes it much easier to find both weight distribution and center of gravity (photo by Michael Lutfy).

Weights and Measures—Start again by weighing the vehicle at all four wheels. You can do it back at the truck scale, or you can use just two individual scales as described in Chapter 7. But, this time, it will be a lot easier with a set of four scales.

Record the overall weight and the front- and rear-wheel weights. Then raise one end—up to 24 inches if possible—with the wheels at the other end still on the scales, and note how much weight is transferred to that end. It doesn't matter which end of the vehicle is raised and which is left on the scales. Use a heavy-duty hydraulic jack or an overhead chain hoist to lift the vehicle. Don't even think about using a bumper jack. It wouldn't be likely to lift the vehicle high enough and, as high as it did go, it wouldn't hold the vehicle safely while you take measurements underneath.

At a truck scale, you would have to measure the overall weight, then drive the vehicle partway off the scale platform, jack up the end that's now clear and record the weight at the wheels still on the scale. That's a time-consuming activity not likely to endear one to the scale operator—or to any truckers waiting in line. You may,

however, be able to go use the scales of a moving company, although they may charge you a small fee.

With only two individual scales, you would have to weigh each end of the vehicle separately. At the opposite end, you would need two blocks the same height as the scales to place under the wheels to assure that the vehicle is level while being weighed. That would mean additional jacking as you swap the scales and blocks, stretching out what is already a tedious procedure.

So, if at all possible, a set of four scales should be used.

Suspension & Tire Deflection—At the end of the vehicle which remains on the scales, the added weight when the other end is raised could deflect the suspension and tires enough to throw off the readings. In addition, the suspension and tires at the raised end could drop slightly and prevent an accurate measurement of just how far that end has been lifted, a figure you'll need for calculating cg height.

To deactivate the suspension at either end, the shock absorbers can be replaced with solid metal rods of the same length, while the tires on the scales can simply be overinflated to minimize deflection.

Full or Empty Fuel Tank—There are differences of opinion as to how much fuel should be aboard the vehicle while it is being weighed to find cg height. Some say that the tank should be full and others that it should be empty. Still others compromise by saying it should be half full—or half empty, depending on your point of view. One of the

A crew member for the Chevrolet Spice IMSA GTP team checks readings on the inevitable "black box," the digital scale of their weighing setup (photo by Michael Lutfy).

With the flip of a switch, the scale can show the weight on any one wheel or the overall weight on all four (photo by Michael Lutfy).

best suggestions I've heard is to go through the whole procedure with the tank empty, then repeat it with the tank full. That will provide the two extremes in cg height that can occur in normal vehicle operation. Generally, on most conventional vehicles, the cg height is slightly lower with a full tank than it is with an empty one.

Necessary Dimensions—Once you know the weight of the vehicle and the amount of weight transferred when one end is raised, there are three dimensions in inches you need. The first is the wheelbase with the vehicle level, which you would most likely already know; the second is the wheelbase at ground level with one end raised; the third, as indicated earlier, is the distance that one end has been raised.

The figures needed for the cg height on our oval track car are all shown in Fig. 8c. To find cg height, multiply the wheelbase with the vehicle level by the wheelbase with one end of the vehicle raised at least 24 inches by the added weight shown on the scales with the one end raised. Then divide the product of that calculation by the distance the one end has been raised multiplied by the overall vehicle weight. Or, stated as an equation:

$$\text{cg Height} = \frac{\text{level wheelbase} \times \text{raised wheelbase} \times \text{added weight on scales}}{\text{distance raised} \times \text{overall weight}}$$

As noted earlier, you would probably already know the level wheelbase of the vehicle and, of course, you could use a yardstick to check how high the one end has been raised.

Measuring Ground Level Wheelbase—To find the wheelbase at ground level with one end of the vehicle raised, you can either measure or calculate it. To measure it, drop a plumb from the bottom of one of the raised tires. Make a chalk mark where the plumb strikes the ground and, with a tape measure, find how far it is from the center of the ground level wheel on the same side of the vehicle.

Calculating Ground Level Wheelbase—To calculate the wheelbase at ground level, note in Fig. 8c that the two wheelbase measurements and the distance that one end of the vehicle has been raised form a right triangle. You would already know the measurements of two sides of that triangle—the 108-inch wheelbase with the vehicle level and the 24-inch distance that one end has been raised. You can find the third side by applying the Theorem of Pythagoras, which states that, in a right triangle, the square of the side opposite the right angle equals the sum of the squares of the other two sides.

The 108-inch wheelbase is the side opposite the right angle, and 108 squared is 11664. The other known side is the 24-inch lift, and 24 squared is 576. The square of the third side is 11664 minus 576, or 11088. The measurement of the third side would be the square root of 11088, or 105.3.

Added Weight—Finally, to find how much weight has been added on the scales, simply find the difference between the reading when the vehicle was level and the reading after the other end was raised. In our example, the first reading was 1410 pounds and the second 1518 pounds, a difference of 108 pounds. (It's simply coincidence in this case that the level wheelbase in inches and the added weight in pounds both happen to be 108.)

The individual wheel weights and the overall vehicle weight are used to compute the center of gravity position with essentially the same mathematical formulas as those in the accompanying text (photo by Michael Lutfy).

CG HEIGHT

Let's plug our figures into the formula and find the cg height of our oval track car:

$$cg\ height = \frac{108 \times 105.3 \times 108}{24 \times 3000} = \frac{1228219.2}{72000}$$

The answer is 17.0586 inches which, of course, can be rounded down to an even 17 inches, a fairly typical figure for the type of car in question. In Fig. 8d, the cg position both lengthways and vertically has been plotted.

As you can see, the position of the cg isn't difficult to calculate. But getting the data needed to find the position, especially vertically, can be difficult and time-consuming.

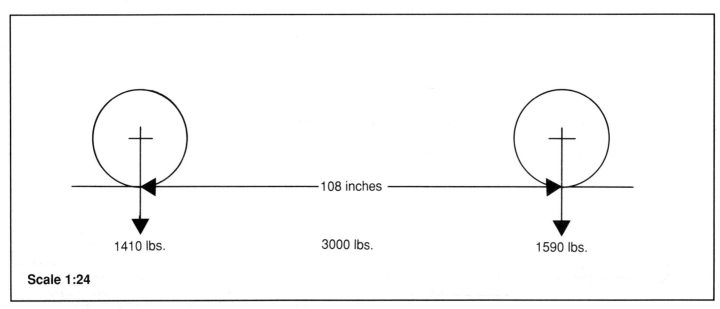

Scale 1:24

Fig. 8a. To find how far the cg is behind the front wheel centers, you need to know the vehicle's rear wheel weight, overall weight and wheelbase.

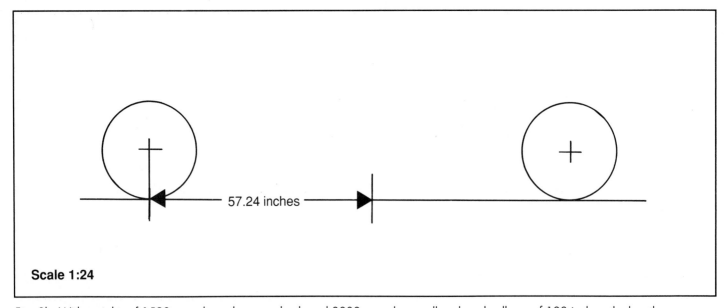

Scale 1:24

Fig. 8b. With weights of 1590 pounds on the rear wheels and 3000 pounds overall and a wheelbase of 108 inches, the lengthways cg position would be 57.24 inches behind the front wheel centers.

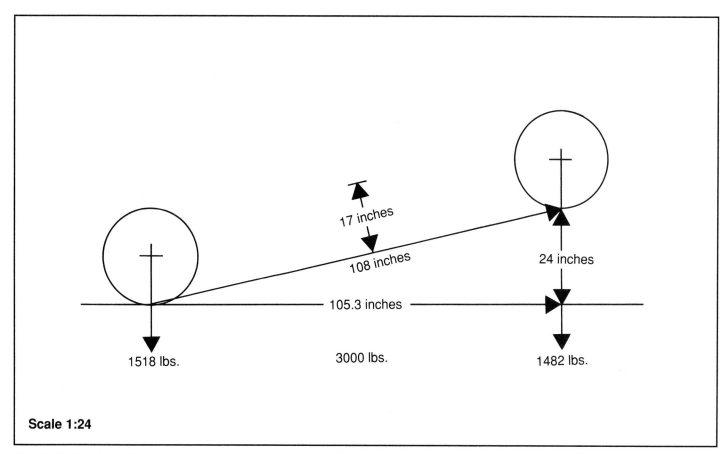

Scale 1:24

Fig. 8c. To find the cg height, it's necessary to raise one end of the car at least two feet, or 24 inches as shown here, and to measure how much weight is added to the scales at the other end. The ground-level wheelbase must also be determined, either by measurement or geometric calculation. Applying the formula described in the text to the measurements in the drawing provides a cg height of approximately 17 inches.

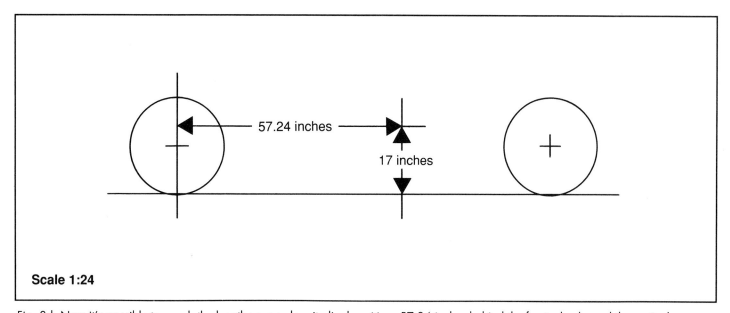

Scale 1:24

Fig. 8d. Now it's possible to graph the lengthways or longitudinal position, 57.24 inches behind the front wheels, and the vertical position, 17 inches. The sideways or lateral position can also be calculated easily, using the weight on the left or right wheels, the weight overall and the vehicle track.

Table 8

FORMULAS FOR CENTER OF GRAVITY

$$\text{cg location behind front wheels} = \frac{\text{rear wheel weight}}{\text{overall weight}} \times \text{wheelbase}$$

$$\text{cg location off-center to heavy side} = \frac{\text{track}}{2} - \left(\frac{\text{weight on light side}}{\text{overall weight}} \right) \times \text{track}$$

$$\text{cg height} = \frac{\text{level wheelbase} \times \text{raised wheelbase} \times \text{added weight on scales}}{\text{distance raised} \times \text{overall weight}}$$

When a high performance car breaks from the line under hard throttle in first gear, the amount of force applied at the drive wheels can be phenomenal. Drag racers need to setup the car carefully for proper launching, otherwise they could smoke the tires and lose the race (photo by Michael Lutfy).

The main rationale given in Chapter 8 for finding the position of the center of gravity is that you need it to calculate weight transfer during acceleration or cornering. Weight transfer, in turn, is critical because it can influence how a vehicle's chassis should be set up.

g FORCE

In order to find weight transfer during a particular maneuver, there are factors needed other than the position of the cg. One of the most important is the g (for gravity) force acting on the vehicle during the maneuver.

Here on earth, a free-falling object will gain speed every second by 32.174 feet per second or, as a physicist would say, it accelerates at 32.174 feet per second *per second*. That can also be written 32.174 feet per second squared. That figure is 1.0 g and is the scientific norm for measuring the acceleration of any moving object, not just one in free fall. When you're in a rapidly accelerating vehicle and you feel as if you're being thrust back into your seat, you're experiencing g force.

To find the g force acting on a car while it's accelerating, you need to know the thrust in pounds being applied by the drive wheels to the road surface. And to find the thrust at the drive wheels, you need to know the torque at the wheels and the rolling radius of the wheels and tires. Here, things begin to get a little complicated.

Drive Wheel Torque—In discussing indicated versus brake engine output in Chapter 5, I pointed out that friction and inertia within the engine cause losses in horsepower and torque between the combustion chambers and the flywheel. Well, friction and inertia make some further claims between the flywheel and the drive wheels, with the transmission and final-drive assembly taking a particular toll on horsepower and torque.

You can find the results of those losses by testing the vehicle on a chassis dynamometer, which measures output at the drive wheels. Or you can simply estimate the losses at about 15 percent, which is what they're likely to be on most modern cars. In other words, the drivetrain should be about 85 percent efficient. A car which has 100 brake horsepower at the flywheel should have about 85 horsepower at the drive wheels.

To find the maximum torque in pounds-feet at the drive wheels, you have to multiply the torque at the flywheel by both the first-gear ratio and the final-drive ratio, and by our 0.85 efficiency factor, or:

$$\text{Drive Wheel Torque} = \frac{\text{flywheel}}{\text{torque}} \times \frac{\text{first}}{\text{gear}} \times \frac{\text{final}}{\text{drive}} \times 0.85$$

Let's demonstrate that with an example, using a late-model Chevrolet Corvette with a 350-cubic-inch engine which has a maximum torque of 330 pounds-feet. The vehicle also has a 5-speed manual transmission with a 2.88:1 first gear and a 3.07:1 final-drive ratio. Applying those figures in the formula, you have:

$$\text{Drive Wheel Torque} = 330 \times 2.88 \times 3.07 \times 0.85$$

That provides a figure of 2480.0688 or, rounded down, 2480 pounds-feet at the wheels. That's right—2480! No wonder it's easy to smoke a 'Vette's rear tires in first gear!

Wheel Thrust—As noted in Chapter 4, torque can be described as a force in pounds applying leverage over a distance in feet; hence our definition of it in pounds-feet.

At the drive wheels, that distance is determined by the tire size or, to be more precise, by the tire's *rolling radius*. That's the vertical measurement from the center of the wheel to the tire's point of contact on the ground. Because the weight of the vehicle flattens the tires slightly, the rolling radius is usually slightly less than the horizontal radius.

To find the thrust in pounds the drive wheels apply to the pavement, divide the torque at the wheels in pounds-feet by the rolling radius in feet:

A Corvette equipped with a 350 cubic-inch engine, which has a maximum torque of 330 lbs-ft., a 2.88 first gear and 3.07 final-drive ratio delivers 2480 lbs-ft. of torque at the wheels. No wonder it's easy to light 'em up! (photo courtesy Corvette Fever Magazine).

$$\text{Wheel Thrust} = \frac{\text{drive wheel torque}}{\text{rolling radius}}$$

Using a yardstick or tape measure, it's easier to get an accurate reading of the rolling radius in inches and convert it to feet, rather than trying to measure it directly in feet. On the 'Vette, suppose the rolling radius is 12.6 inches. To convert that to feet, divide by 12, giving you a figure of 1.05 to divide into the drive wheel torque:

$$\text{Wheel Thrust} = \frac{2480}{1.05}$$

The thrust at the drive wheels is 2361.9048 or, rounded up, 2362 pounds.

Calculating g Force—To find the *g* force during acceleration, you simply divide the thrust in pounds by the vehicle weight, or:

$$g \text{ Force} = \frac{\text{wheel thrust}}{\text{weight}}$$

Given a weight of 3292 pounds for the Corvette, you would divide that into the thrust figure of 2362 pounds:

$$g \text{ Force} = \frac{2362}{3292}$$

The 'Vette's potential rate of acceleration would be 0.717497 or, rounded down, 0.72 g.

We already know that 1.0 *g* equals 32.174 feet per second per second. Multiplying that by 0.72, the 'Vette's potential acceleration could also be expressed as 23.16528 or rounded off, 23 feet per second per second.

Those are strictly theoretical figures, though, that don't take into account such variables as rolling resistance or aerodynamic drag. Nonetheless, if you know the maximum *g* forces for a variety of vehicles, they do have comparative value.

WEIGHT TRANSFER

Weight transfer is especially important in drag racing. As a car breaks from the line at the start of a 1/4-mile run, the weight that shifts momentarily to the rear will apply force to the drive wheels that should improve traction; the greater the weight transferred, the better the bite.

To find a vehicle's maximum weight transfer during acceleration, multiply the overall weight by the height of the cg, divide that by the wheelbase, and then multiply the result by the *g* force, or:

When a vehicle is accelerating rapidly—as in a drag race—weight is transferred rearward, improving traction on the rear wheels (photo by Michael Lutfy).

$$\text{Weight Transfer} = \frac{\text{weight} \times \text{cg height}}{\text{wheelbase}} \times g$$

$$\text{Weight Transfer} = \frac{3292 \times 24}{96.2} \times 0.72$$

You already have a weight of 3292 pounds and a g force of 0.72 for the Corvette. Its cg height would be approximately 18 inches and its wheelbase 96.2 inches. So, to apply the formula:

$$\text{Weight Transfer} = \frac{3292 \times 18}{96.2} \times 0.72$$

The maximum potential weight transferred to the rear wheels during acceleration would be 443.49605 or, rounded down, 443 pounds.

To show the effect of cg height on weight transfer, suppose you wanted to rebuild the Corvette for drag racing. You were able to jack it up enough that the cg was raised 6.0 inches to a height of 24 inches while, for the sake of simplicity, the other critical specs were kept the same:

That would increase weight transfer to 590.10124 or, rounded down, 590 pounds, a gain of over 33 percent or one-third!

In actual practice, if the car were being prepared for the drags, the engine would have been modified for higher output and a numerically higher final-drive ratio installed. With greater torque and stronger gearing, the potential g force would have been raised and added still further to the amount of potential weight transfer.

That explains why many drag cars, in classes which permit it, are built as high off the ground as they are.

LATERAL ACCELERATION

In most forms of motorsports other than drag racing, weight transfer is kept as low as possible for steady, consistent handling. In both road and oval track racing, for

Unlike drag racing, weight transfer is kept as low as possible in most other forms of motorsports for steady, consistent handling. A low center of gravity, and with it, minimum weight transfer are desirable for maximum cornering ability (photo by Michael Lutfy).

example, a low center of gravity and, with it, minimum weight transfer are desirable for cornering stability.

During straightaway acceleration on a drag strip, as you've seen, the key to the force being applied to the vehicle is the thrust in pounds at the drive wheels. During cornering on an oval track or road course, the key is the *g* force acting on the vehicle and attempting to push it sideways as it goes around the turn.

This sideways *g* force is called *lateral acceleration*, a factor you'll need to calculate lateral weight transfer.

Calculating—To calculate lateral acceleration, you'll need two factors that can only be determined by testing the vehicle on a skid pad. As Fred Puhn explains in HPBooks' *How to Make Your Car Handle*:

> "A skid pad is a flat piece of pavement with a circle painted on it. The car is driven around the circle, keeping the center of the car right on the line. By measuring the time it takes to make one lap of the circle, the lateral acceleration can be computed. To do this, you need to know the radius of the circle and the time for one lap at maximum speed."

The radius of the circle should be in feet and the time for one lap in seconds. The raw formula for the lateral acceleration in feet per second per second is the square of 2.0 times *pi*, multiplied by the radius divided by the square of the time, or:

$$\text{Lateral Acceleration} = (2.0 \times pi)^2 \times \frac{\text{radius}}{\text{time}^2}$$

To find lateral acceleration directly in *g* force, the value of 1.0 *g* in feet per second per second—which, of course, is 32.174—can be plugged into the formula:

$$\text{Lateral Acceleration} = \frac{(2.0 \times pi)^2}{32.174} \times \frac{\text{radius}}{\text{time}^2}$$

The figures involving *pi* and *g* can be reduced to a single constant:

$$\frac{(2.0 \times pi)^2}{32.174} = \frac{6.2831853^2}{32.174} = \frac{39.478418}{32.174}$$

During fast cornering, lateral acceleration tends to force a vehicle sideways, out of the turn. To find lateral acceleration in g force, the car must be timed in seconds at the limits of its adhesion around a skid pad of known radius (photo courtesy Corvette Fever Magazine).

Which works out to 1.2270286 or, rounded down, 1.227. Now the formula for lateral acceleration in g's can be simplified to:

$$\text{Lateral Acceleration} = 1.227 \times \frac{\text{radius}}{\text{time}^2}$$

Example—Let's suppose you test the Corvette on a skid pad with a radius of 150 feet and it turns a lap in 14.5 seconds:

$$\text{Lateral Acceleration} = 1.227 \times \frac{150}{14.5^2}$$

$$\text{Lateral Acceleration} = 1.227 \times \frac{150}{210.25}$$

The result would be a lateral acceleration figure of 0.8753864 or, rounded down, 0.875 g.

LATERAL WEIGHT TRANSFER

To find the weight transfer during cornering, the formula is essentially the same as the one for weight transfer during acceleration, except that the vehicle's wheel track is used instead of its wheelbase:

$$\text{Lateral Weight Transfer} = \frac{\text{weight} \times \text{cg height}}{\text{wheel track}} \times g$$

You already know the weight of the Corvette in our ongoing example is 3292 pounds and its cg height is 18 inches. The 'Vette has a front track of 59.6 inches and a rear track of 60.4 inches for an average of an even 60 inches. So to find its sideways weight transfer:

$$\text{Lateral Weight Transfer} = \frac{3292 \times 18}{60} \times 0.875$$

The answer is 864.15 pounds.

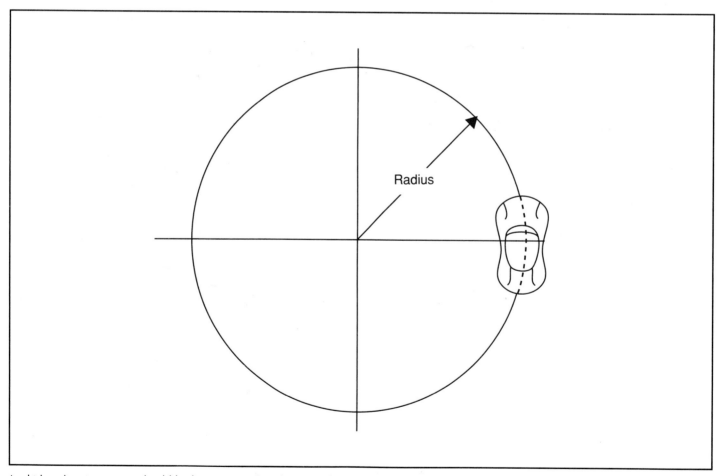

In skid pad testing, a car should be kept centered on the circumference of the skid pad circle. This diagram is adapted from HPBooks' How to Make Your Car Handle by Fred Puhn, an excellent reference on chassis engineering for high-performance cars.

You can reduce the amount of weight transferred in a turn by decreasing the weight and/or the cg height and/or by increasing the track—none of which is particularly easy to do.

CENTRIFUGAL FORCE

Given the vehicle's weight and the *g* force determined in a skid pad test, you can also find the centrifugal force in pounds acting on the vehicle while cornering with a simple formula:

$$\text{Centrifugal Force} = \text{weight} \times g$$

For the Corvette, you'd multiply 3292 x 0.875, which would work out to a centrifugal force of 2880.5 pounds. That, however, is mainly a point of academic interest, and not of the same practical significance as the *g* force or weight transfer values.

Table 9

FORMULAS FOR g FORCE & WEIGHT TRANSFER

$$\text{drive wheel torque} = \frac{\text{flywheel}}{\text{torque}} \times \frac{\text{first}}{\text{gear}} \times \frac{\text{final}}{\text{drive}} \times 0.85$$

$$\text{wheel thrust} = \frac{\text{drive wheel torque}}{\text{rolling radius}}$$

$$g = \frac{\text{wheel thrust}}{\text{weight}}$$

$$\text{weight transfer} = \frac{\text{weight x cg height}}{\text{wheelbase}} \times g$$

$$\text{lateral acceleration} = 1.227 \times \frac{\text{radius}}{\text{time}^2}$$

$$\text{lateral weight transfer} = \frac{\text{weight x cg height}}{\text{wheel track}} \times g$$

$$\text{centrifugal force} = \text{weight} \times g$$

Shift Points

Even the most humble doorslammer can be driven to quicker ETs and faster terminal speeds if the driver knows when to shift for optimum performance. But that takes homework! (photo by Michael Lutfy).

To get the best acceleration out of a high-performance vehicle during shifts, hot rodders say you should stay "on the cam." What they mean is that you should keep the engine within an rpm range where the transmission is delivering optimum torque before and after each shift.

The first step toward finding what that range might be is to use a dyno chart. So, once more, let's turn to the dyno chart in Fig. 10a, which is the same one used in Chapter 4. In order to calculate shift points, though, you'll be primarily concerned with rpm and torque, not horsepower, because it's torque that accelerates an automobile.

Next, you need to know the loss or gain in rpm when you shift from one gear to another.

Example—Suppose you have a Chevrolet powered by the modified 350-cubic-inch engine with output specifications shown in Fig. 10a, with a Warner T-10 4-speed gearbox which has ratios of 2.20 in 1st, 1.66 in 2nd, 1.31 in 3rd and direct 1.00 drive in 4th. When you drag race, you shift at 6000 rpm. How much rpm do you lose during the shift from, say, 1st to 2nd?

You can find out by dividing the ratio in 1st gear into the

ratio in 2nd. The result will be a percentage which, when multiplied by the rpm in 1st, will provide the equivalent rpm in 2nd. Or, expressed as an equation:

$$\text{RPM After Shift} = \frac{\text{ratio shift into}}{\text{ratio shift from}} \times \text{rpm before shift}$$

With the Warner T-10, divide the 2.20 1st-gear ratio into the 1.66 2nd-gear ratio and multiply by 6000:

$$\text{RPM After Shift} = \frac{1.66}{2.20} \times 6000 = 0.7545455 \times 6000$$

The engine speed in 2nd will be 4527 rpm. Subtracting that figure from 6000, you'll find you have a drop of 1473 rpm or about 25 percent. Obviously, you can also apply the formula to shifts from 2nd to 3rd and from 3rd to 4th. From 2nd to 3rd, the engine speed will drop from 6000 to 4735 rpm, a loss of 1265 rpm or 21 percent; from 3rd to 4th, it will fall from 6000 to 4580 rpm, losing 1420 rpm or 27 percent.

DRIVESHAFT TORQUE

But, given the torque characteristics shown on the dyno chart, is 6000 rpm the best point at which to upshift? To answer that question, you need to know the driveshaft torque being delivered to the rear wheels before and after each shift. That's simply a matter of multiplying the brake torque at the flywheel by the transmission ratio, or:

$$\text{Driveshaft Torque} = \text{flywheel torque} \times \text{transmission ratio}$$

According to Fig. 10a, you have 300 pounds-feet of torque at the flywheel at 6000 rpm. With a 1st gear ratio of 2.20, that becomes 660 pounds-feet being delivered from the transmission via the driveshaft to the drive wheels.

What about the friction mentioned in Chapter 9? Won't the output from the transmission to the driveshaft be slightly less than the input to the transmission from the flywheel? Yes, it will be. But it doesn't really affect the comparative validity of our driveshaft torque figures so, to simplify your calculations, you needn't take it into account here.

All right, you have 660 pounds-feet of driveshaft torque at 6000 rpm in 1st gear. When you shift into 2nd, the rpm

RPM	LB-FT	BHP
3000	340	194
3200	340	207
3400	345	223
3600	345	236
3800	350	253
4000	350	267
4200	340	272
4400	335	281
4600	330	289
4800	325	297
5000	315	300
5200	310	307
5400	305	314
5600	305	325
5800	305	337
6000	300	343
6200	280	331
6400	255	311
6600	240	302
6800	190	246
7000	160	213

Figure 10a. This chart shows the torque and horsepower for a modified 350 cubic-inch Chevy at intervals of 200 rpm from 3000 to 7000 rpm as measured on a dyno.

drops to 4527. According to Fig. 10a, the engine has 330 pounds-feet of torque at 4600 rpm. Multiply that by a 2nd gear ratio of 1.66 and the result would be 548 pounds-feet of driveshaft torque. During the shift from 1st to 2nd, you've lost 112 pounds-feet of driveshaft torque or about 17 percent. A 17-percent drop in torque during an upshift doesn't sound like a good way to win a drag race.

IDEAL SHIFT POINTS

To find the ideal shift points for this particular combination of Chevy engine and Warner gearbox, the two formulas discussed in this chapter were used to construct the charts shown in Fig. 10b on page 75. There's a separate chart for each shift, from 1st to 2nd, 2nd to 3rd, and 3rd to 4th, and shift points are shown at 200-rpm intervals from 6000 to 7000 rpm.

The rpm immediately before the shift is in the 1st column. In the 2nd is the brake flywheel torque at that rpm, as shown on the engine dyno chart, and in the 3rd is the driveshaft torque at that rpm.

The 4th column displays the rpm immediately after the shift. The 5th shows the flywheel torque at the rpm on the dyno chart closest to the rpm after the shift, and the 6th the driveshaft torque after the shift. (Note that in 4th gear, which is direct drive, there's no difference between the flywheel and driveshaft torque figures.)

In the 7th column is the change, minus or plus, in driveshaft torque after the shift. After each of the three shifts, the least change in the 7th column occurs at 6600 rpm. During the shift from 1st to 2nd, there is a loss of only 5.0 pounds-feet at that engine speed; from 2nd to 3rd, a loss of 8.0 pounds-feet; and from 3rd to 4th, a gain of 1.0 pound-foot.

Below 6600 rpm, there are much greater drops in torque after the shift. Above 6600, there are gains in torque after the shift but at the price of losing torque and, with it, momentum before the shift. So 6600 rpm seems to be the best shift point across the board.

There's a drag racer's rule of thumb that the best shift point is at an engine speed 10 percent beyond the horsepower peak. The dyno chart shows the horsepower peak at 6000 rpm and, of course, 6600 rpm is exactly 10 percent beyond that. However, that should be regarded as

an interesting coincidence and not conclusive proof of the rule of thumb.

Downshifts—The formulas can also be used to calculate changes in rpm and torque during downshifts. Let's say you're running a T-10 equipped sports car in a road race. You're heading toward a turn at 4000 rpm in 3rd gear. How high should the engine be revved for a downshift to 2nd?

Divide 1.31 into 1.66 and multiply by 4000. The engine should be turning a theoretical 5069 rpm as you go from 4th into 3rd; "theoretical" because not even an Andretti or Unser can coordinate a tach reading and throttle pressure closely enough to achieve that precise an engine speed! You would gain a theoretical 1069 rpm, an increase of approximately 27 percent. During that shift—using the torque figures from the Fig. 10a at 4000 and 5000 rpm—there'd be a gain of 64 pounds-feet in driveshaft torque. Here, too, it would be instructive to develop charts showing the changes at different shift points.

These formulas aren't complicated but, obviously, you could spend a lot of time with them, calculating different combinations of torque, rpm and gearing. And those calculations might just win a race or two.

RPM	brake lb-ft	shaft lb-ft	RPM	brake lb-ft	shaft lb-ft	loss/ gain
Shift from 2.20 1st to 1.66 2nd						
6000	300	660	4527	330	548	-112
6200	280	616	4678	330	548	-68
6400	255	561	4829	325	540	-21
6600	240	528	4980	315	523	-5
6800	190	418	5131	310	515	+97
7000	160	352	5282	310	515	+165
Shift from 1.66 2nd to 1.31 3rd						
6000	300	498	4735	325	426	-72
6200	280	465	4893	05	400	+85
6400	255	423	5051	315	413	-10
6600	240	398	5200	406	406	-8
6800	190	315	5366	305	400	+85
7000	160	266	5524	305	400	+134
Shift from 1.31 3rd to 1.00 4th						
6000	300	393	4580	330	330	-63
6200	280	367	4753	325	325	-42
6400	255	334	4885	325	325	-9
6600	240	314	5038	315	315	+1
6800	190	249	5191	310	310	+61
7000	160	210	5344	305	305	+95

Figure 10b. By using engine dyno figures and gear ratios to plot driveshaft torque before and after upshifts at different rpm, it's possible to determine the shift point at which the loss or gain in torque is lowest.

Table 10

FORMULAS FOR SHIFT POINTS

$$\text{rpm after shift} = \frac{\text{ratio shift into}}{\text{ratio shift from}} \times \text{rpm before shift}$$

$$\text{driveshaft torque} = \text{flywheel torque} \times \text{transmission ratio}$$

Quarter-Mile ET & MPH

The best way to find a car's quarter-mile elapsed time and terminal speed is to run it at the drag strip. However, Patrick Hale's *Quarter* and *Quarter jr.* computer programs can predict quarter-mile performance with surprising accuracy. Even his simplified formulas for ET and mph can provide useful comparative data for different combinations of power and weight (photo by Michael Lutfy).

COMPUTER PROGRAMMING

In 1986, Patrick Hale, a drag racer, engineer and computer programmer, combined his interests and developed computer software that predicts how a car should perform in the quarter mile. Called *Quarter*, the program predicts elapsed times and speeds accurately, because it takes into account practically every mathematical variable that can affect acceleration, from gearing and shift points to drag coefficient and polar moment of inertia. It also shows what modifications will help a car's performance and what will hurt it—before the owner spends cold, hard cash for actual parts. All you need is an IBM PC or compatible or an Apple IIC.

For some racers, the program was too good. It required data that those running on limited budgets simply didn't have and couldn't get. So, a year later, Hale introduced a simpler, lower-priced program not only for the IBM and Apple but for the Commodore 64, too. Called *Quarter jr.*, it focuses on readily available specs and provides its own built-in estimates of the more esoteric details.

At either programming level, the two variables that are

the most important are *horsepower* and *weight*. The higher the power and/or the the lower the weight, the faster and quicker the car will go.

In the course of developing his computer programs, Hale came up with formulas for computing quarter-mile elapsed time or ET and terminal speed in mph from just power and weight. Because they don't consider any other variables, these formulas can't and don't predict performance as precisely as *Quarter* or *Quarter jr*. But they do provide useful—and comparative—estimates.

ELAPSED TIME

The formula for ET involves the cube root of the weight-to-power ratio multiplied by a constant of 5.825, while the formula for terminal mph calls for the cube root of the power-to weight ratio multiplied by a constant of 234. Note an important distinction here: For ET, you want weight-to-power, i.e., pounds per horsepower. For mph, you want power-to-weight, i.e., horsepower per pound.

$$ET = \sqrt[3]{\frac{weight}{hp}} \times 5.825$$

Here's the ET formula in proper mathematical form: Suppose you have a Corvette that weighs 3440 pounds,

Quarter jr. is available in versions for the Apple II, Commodore 64, IBM PC and compatibles. For further details about the program, write Patrick Hale, Racing Systems Analysis, P.O. Box 7676, Phoenix, Arizona or call (602) 241-1301.

complete with fuel and the driver aboard, with a 245-hp engine. According to the formula, its quarter-mile ET should be 14.05 seconds. *Road & Track* once tested just such a combination and posted an elapsed time of 14.6 seconds. Without considering any other variables, Hale's formula has come within 4.0 percent of an actual test run.

Power or Weight from ET—From the formula for ET, formulas can be derived to find either power or weight, when the other is known. To find how much hp would be needed to propel a car of a given weight to a given ET, divide the weight by the cube of the ET divided by 5.825:

$$HP = \frac{weight}{(ET/5.825)^3}$$

For the 3440-pound 'Vette to post an ET of 14.05 seconds, it would need—surprise!—245 hp.

If the unknown were the weight, you could find it with the cube of the ET divided by 5.825, multiplied by the hp:

$$Weight = (ET/5.825)^3 \times hp$$

And how much should the 245-hp, 14.05-second 'Vette weigh? The answer is 3438 pounds. (The calculator lost a couple of pounds in rounding errors.)

MILES-PER-HOUR

Patrick Hale's formula for mph at the end of a quarter-mile acceleration run is:

$$MPH \sqrt[3]{\frac{hp}{weight}} \times 234$$

For the Corvette, the speed would be 97 mph. *Road & Track's* test figure was 95.5 mph. So, this time, the formula's error is less than 1.6 percent!

Power or Weight from MPH—Again, formulas can be derived to find either power or weight. To find the hp, the cube of the mph divided by 234 should be multiplied by the weight:

$$HP = (mph/234)^3 \times weight$$

To propel the 3440-pound 'Vette to 97 mph in the quarter-mile, you'd need—yes—245 hp.

To find the weight, the cube of 234 divided by the mph should be multiplied by the hp:

$$\text{Weight} = (234/\text{mph})^3 \times \text{hp}$$

Here, the weight works out to 3439.5 pounds, only 1/2 pound off.

REALISTIC INPUT = REALISTIC OUTPUT

As I said earlier, these formulas aren't as precise as Hale's computer programs, and they won't show the effects of any modifications in areas other than power and weight. But, considering their simplicity and how easy they are to work, they come remarkably close to real-life figures.

How close depends on the accuracy of the *input*, and that can be a problem. Drino Miller, a race car builder and driver I know, once remarked that he'd tell anyone whatever they wanted to know about any of his cars except two things, their power and their weight, because, in his words, "Everybody lies about those."

He had a point. Exaggerated claims about power and weight are among the most common tactics racers use in their constant efforts to psych out one another.

It was in the interest of accuracy that I used *Road & Track* test figures to demonstrate Hale's quarter-mile formulas. *Road & Track* actually weighs its test vehicles. while many other magazines simply print curb weights provided by the auto makers, and those are often highly optimistic

But, as I've tried to show, with realistic input, the formulas can provide surprisingly realistic output.

GEARING FOR QUARTER-MILE SPEED

One of the tacit assumptions of Patrick Hale's formulas is that the vehicle is properly geared and that, of course, may or may not be true. It's possible that the gearing might

This is an actual test using Quarter jr. for a vehicle called "etracer." The program takes numerous variables into account in estimating ET and mph, including weather conditions, gearing, tire size and body design. But the two most important factors for "etracer" are its engine output of 250 horsepower and weight of 3500 pounds.

be too low—or too high—to enable the car to reach the quarter-mile speed indicated by the formula.

However, once the potential speed has been calculated, there's a formula from another source for determining the optimum overall gearing. It's from Larry Shepard of *Mopar Performance* and, for a car with a manual gearbox, it is:

$$\text{Overall Gear Ratio} = \frac{\text{tire diameter}}{340} \times \frac{\text{rpm}}{\text{mph}}$$

The Corvette discussed earlier has 275/40ZR17 tires, which would have a diameter of 25.66 inches. (How we found the diameter is explained in Chapter 14, page 96.) The engine could be revved easily to 5500 rpm, and Hale's formula predicted a speed at the end of the quarter mile of 97 mph. Plugging the necessary figures into Shepard's formula:

$$\text{Overall Gear Ratio} = \frac{25.66}{340} \times \frac{5500}{97}$$

$$\text{Overall Gear Ratio} = 0.0754706 \times 56.7$$

The recommended overall gear ratio would be 4.279183, or about 4.28.

How does that compare with the Corvette's actual gearing? Well, it has a 3.07 final drive but, as it cleared the end of the quarter, it was still in the 3rd of its 4 gears, with

```
                    QUARTER jr - Version 1.02
F1 - Refresh Input          F3 - Retrieve File        F5 - Directory
F2 - Execute QUARTER jr      F4 - Save File            F6 - Exit QUARTER jr
File: etracer.dat    Title: Test Case for QUARTER jr Version 1.0
      Time     Distance      MPH    Acceleration   Gear      RPM
      0.00        0          0.0    1.18 (s)        1       3,500
    .211/0.00   Rollout      6.1    1.14            1       3,500
      1.11       30         27.5    0.65            1       3,500
      1.76       60         35.2    0.49            1       4,300
      2.00       73         37.7    0.46            1       4,600
      2.39       95         41.0    0.37            1       5,000
      2.47       100        41.8    0.42            2       3,500
      4.00       200        54.2    0.34            2       3,930
      6.00       380        67.3    0.25            2       4,880
      6.35       423        69.1    0.23            2       5,000
      6.45       433        69.6    0.24            3       3,500
      8.00       600        77.3    0.22            3       3,770
      8.52       660        79.8    0.21            3       3,890
     10.00       840        86.5    0.20            3       4,210
     11.22      1,000       91.6    0.19            3       4,450
     12.00      1,106       94.7    0.18            3       4,590
     13.50      1,320       99.9    0.15            3       4,840

        ELAPSED TIME  13.50 SEC            SPEED   99.9 MPH
             Press Desired Function Key:
```

Quarter jr. provides a detailed analysis of how "etracer" should perform throughout the quarter-mile, but the bottom line is an ET of 13.50 seconds and speed of 99.9 mph. Applying the car's power and weight in Hale's simplified formulas described in the accompanying text, the ET would be 14 seconds and the speed 97.089 mph.

a ratio of 1.34. Multiplying 3.07 by 1.34, the overall ratio is 4.11. That's just slightly more than 4.0 percent off the formula's recommendation!

With an automatic transmission, the constant 335 replaces 340:

$$\text{Overall Gear Ratio} = \frac{\text{tire diameter}}{335} \times \frac{\text{rpm}}{\text{mph}}$$

Had the 'Vette been equipped with a Turbo Hydra-Matic, the recommended overall gear ratio to reach a quarter-mile terminal speed of 97 mph would've been 4.3431297 or, rounded off, 4.34.

CONSTANT SOURCE

Where do formulas like these come from? In particular, where did Patrick Hale get his constants 5.825 and 234 and Larry Shepard his 340 and 335? The answer is that the constants were derived empirically. That's a fancy way of saying by trial and error—a lot of trial and error!

According to the formula in the text, a Corvette weighing 3440 pounds would need 245 horsepower to post an ET of 14.05 seconds (photo courtesy Corvette Fever).

TABLE 11

FORMULAS FOR QUARTER-MILE ET & MPH

$$ET = \sqrt[3]{\dfrac{weight}{hp}} \times 5.825$$

$$hp = \dfrac{weight}{(ET/5.825)^3}$$

$$weight = (ET/5.825)^3 \times hp$$

$$mph = \sqrt[3]{\dfrac{hp}{weight}} \times 234$$

$$hp = (mph/234)^3 \times weight$$

$$weight = (234/mph)^3 \times hp$$

$$\begin{array}{c} \text{overall gear ratio =} \\ \text{(manual transmission)} \end{array} \quad \dfrac{tire\ diameter}{340\ mph} \times \dfrac{rpm}{mph}$$

$$\begin{array}{c} \text{overall gear ratio =} \\ \text{(automatic transmission)} \end{array} \quad \dfrac{tire\ diameter}{335\ mph} \times \dfrac{rpm}{mph}$$

Instrument Error 12

The tachometer can be tested for accuracy on an electronic diagnostic machine at the neighborhood garage, while the speedometer and odometer can be checked against measured miles on the highway. These are the analog instruments in a Chevy Cavalier Z24.

When a car's performance is measured at a drag strip, oval track or road course, it's normally done with the racing facility's timing equipment, using instruments that are finely calibrated for accuracy. After all, races are often won or lost by hundredths and sometimes even thousandths of a second.

But when a car's performance is measured on the street or highway, the driver usually has to rely on the vehicle's own instruments—specifically its tachometer, speedometer and odometer—and they're not likely to be all that finely calibrated!

TESTING INSTRUMENTS

The tachometer is the easiest to check for accuracy. At your neighborhood garage, there's likely to be an electronic diagnostic machine that can be hooked up to show engine rpm, and you can simply ask the mechanic to compare your tachometer's reading with the machine's.

The speedometer and odometer are more problematical and, to check them, you may have to take the car to a shop specializing in automotive instruments. Or, if you belong to an auto club, such as an affiliate of the American Automobile Association, it may have facilities where you can have both the speedometer and odometer tested.

SPEEDOMETER CHECK

It's also possible to check the speedometer and odometer on your own. For the speedometer, you'll need a stopwatch, a level stretch of highway with a marked measured mile, and either a cruise control or a well-disciplined right foot.

In some states, there are posted speedometer checks along major highways, with 5 or 10 marked miles. But even where there aren't such posted checks, there are often less conspicuous mile markers along the roadside.

With the stopwatch, time the vehicle over the marked

Digital speedometers, such as this one showing a big "O" on the instrument panel of an Oldsmobile Bravada, can be checked on the highway in the same way as their analog counterparts.

mile at a steady indicated speed. To find your actual speed in mph, you divide the number of seconds it takes you to drive the mile into 3600, the number of seconds in an hour:

$$\text{Actual MPH} = \frac{3600}{\text{seconds per mile}}$$

If the indicated speed is more than the actual speed, the speedometer is fast. If the indicated speed is less than the actual speed, the speedometer is slow.

Example—If you drive a measured mile at an indicated 50 mph and it takes 72 seconds:

$$\text{Actual MPH} = \frac{3600}{72}$$

The actual speed would be the same as the indicated, 50 mph. At that speed, at least, the speedometer would be correct. In other words, there is no error.

Now, let's suppose the measured mile at an indicated 50 mph takes 75 seconds:

$$\text{Actual MPH} = \frac{3600}{75}$$

The actual speed would be 48 mph. That's 2 mph less than the indicated speed, so the speedometer is fast.

Okay, suppose the time was only 70.5 seconds:

$$\text{Actual MPH} = \frac{3600}{70.5}$$

According to the calculator, that would be 51.06383 mph, which should be rounded down to 51.06. That's 1.06 mph more than the indicated 50, so the speedometer is slow.

Speedometer Error Percentage—If the speedometer error is substantial, you should have the speedometer gear replaced with one that will provide more accurate readings. To choose the new gear, the speedometer shop will need to know the percentage of error with the present gear.

To determine the percentage, start with the difference between the indicated and actual mph by subtracting the smaller figure from the larger. Then divide that figure by the actual mph, and multiply the resulting decimal by 100 to convert it to percent:

$$\text{Percent Error} = \frac{\text{actual speed–indicated speed}}{\text{actual speed}} \times 100$$

When you went an actual 48 mph at an indicated 50 mph, the difference was 2:

$$\text{Percent Error} = \frac{2}{48} \times 100$$

At an indicated 50 mph, the speedometer was 4.17 percent fast.

In the example where you went an actual 51.06 mph at an indicated 50 mph, the difference was 1.06:

$$\text{Percent Error} = \frac{1.06}{51.06} \times 100$$

At an indicated 50 mph, the speedometer was 2.08 percent slow.

ODOMETER ERROR

To find the odometer error, it's desirable to use more than 1 measured mile—5 miles at least, 10 if possible. The greater the distance, the more accurate the check will be. This time, though, you don't have to drive the vehicle at a steady speed or to time it with the stopwatch.

Note the odometer reading at the beginning of the measured distance and again at the end. To find the indicated distance, subtract the first figure from the second, or:

$$\text{Indicated Distance} = \frac{\text{Reading}}{\text{at finish}} - \frac{\text{Reading}}{\text{at start}}$$

On an analog (as opposed to digital) odometer, pay particular attention to the reel at the right end, indicating tenths of a mile. If the number isn't centered right in line with the odometer's other numbers, interpolate the reading out to hundredths. For example, if at the start or finish of a run, the reel's display is midway between 2 and 3, it's

On highways where there are no posted speedometer checks, there are often less conspicuous mile markers along the roadside, such as surveyors' marks, but it may take some cruising to find them.

showing 0.25 mile. If it's 2/3 of the way between those two numbers, it's showing approximately 0.27 mile.

If the indicated distance shown on the odometer is more than the measured distance, the odometer is fast. If the indicated distance is less, the odometer is slow.

Odometer Error Percentage—To find the percentage of odometer error, use the same formula you did for speedometer error, but with actual and indicated distances rather than speeds, or:

$$\text{Percent Error} = \frac{\text{difference between actual and indicated distances}}{\text{actual distance}} \times 100$$

Suppose that over a measured 5-mile stretch, the odometer actually records 5.15 miles. To find the percentage of error, divide the 0.15 difference between the actual and indicated figures by the actual figure and, of course, multiply by 100:

$$\text{Percent Error} = \frac{0.15}{5} \times 100$$

The odometer is 3.0 percent fast.

INCONSISTENCIES

In the examples of how to find speedometer error, the figures used were for an indicated speed of 50 mph. The percentage of error won't necessarily be the same at other speeds and a professional speedometer check will cover a broad range of speeds.

Similarly, the percentage of odometer error won't necessarily correspond with the speedometer error. If one reads fast, the other probably reads fast, too, but not always to the same degree.

A real world example of such discrepancies is shown in a chart in Fig. 12a. When the final drive gears in my own car were replaced by gears with a numerically higher ratio, the speedometer and odometer were thrown way off. The figures in the first two columns of the chart are the actual and indicated readings recorded at an auto club facility. The figures in the other two columns show the differences between the actual and indicated readings and the percentages of error at the various speeds checked.

At the bottom of the chart are the figures for the odometer. Note that its percentage of error is considerably less than any of the percentages for the speedometer.

Finally, to answer an obvious question: yes, the speedometer drive gear has since been replaced with one that's brought the readings back down closer to reality!

Speedometer			
ind.speed	act.speed	dif. error	percent
25	20	5	25
30	25	5	20
35	28	7	25
40	33	7	21.2
45	38	7	18.4
50	42	8	19
55	46	9	19.6
60	50	10	20
65	54	11	20.3
70	58	12	20.6
75	62	13	20.9
80	66	14	21.2
Odometer			
ind. distance	act. distance	dif. error	percent
100	88	12	13.6

Fig. 12a

Table 12

FORMULAS FOR INSTRUMENT ERROR

$$\text{actual mph} = \frac{3600}{\text{seconds per mile}}$$

$$\text{speedo error percent} = \frac{\text{difference between actual and indicated speeds}}{\text{actual speed}} \times 100$$

$$\text{indicated distance} = \frac{\text{odometer reading}}{\text{at finish}} - \frac{\text{odometer reading}}{\text{at start}}$$

$$\text{odo error percent} = \frac{\text{difference between actual and indicated distances}}{\text{actual distance}} \times 100$$

MPH, RPM, Gears & Tires 13

The formulas for engine speed in rpm, vehicle speed in mph, overall gear ratio and tire diameter can be useful for analyzing the behavior on the road, track or strip of high-performance cars like this Ford Mustang.

There are four significant, interrelated specifications—speed in miles per hour or *mph*, engine revolutions per minute or *rpm*, overall gear ratio and tire diameter in inches. Given any three of these, it's possible to determine what the fourth is—or should be.

Determining the values for these four areas can be useful for analyzing the behavior of high performance cars on the road, track or strip.

MILES PER HOUR

The raw formula for finding vehicle mph involves multiplying engine rpm by 60 (the number of minutes in an hour) by *pi* times the tire diameter (which provides the circumference of the tire in inches) and then dividing by the overall gear ratio times 63360 (the number of inches in a mile), or:

$$MPH = \frac{rpm \times 60 \times pi \times tire\ diameter}{gear\ ratio \times 63360}$$

$$MPH = \frac{143,000}{3466.848}$$

The constants in the numerator, 60 and *pi* or 3.1415927, can be multiplied together to become 188.49556. That figure, in turn, can be divided into both the numerator and denominator, eliminating it from the numerator and reducing the constant in the denominator to 336.13524. That can be rounded down to 336, and the formula becomes a much more manageable:

$$MPH = \frac{rpm \times tire\ diameter}{gear\ ratio \times 336}$$

Example—To demonstrate that formula, let's suppose you have a Ford Mustang with a 5.0-liter V-8 and a 5-speed transmission with ratios of 3.35 in 1st, 1.93 in 2nd, 1.29 in 3rd, direct 1.00 in 4th and overdrive 0.68 in 5th. The final-drive ratio is 3.08 and the tires have a diameter of 26 inches.

In the quarter-mile, you run through 1st, 2nd and 3rd, shifting at 5500 and you want to know what speed you reach in each gear. For 1st gear, multiply 5500 by 26 and then divide by the 1st gear ratio of 3.35 times the final drive ratio of 3.08 times the constant 336:

$$MPH = \frac{5500 \times 26}{3.35 \times 3.08 \times 336}$$

By completing the division, you'll find that at 5500 rpm in 1st gear, the Mustang will be going 41.247842 mph or, rounded up, 41.25 mph. Similarly, the formula will show that at 5500 rpm in 2nd, the little Ford will be doing 71.60 mph, and in 3rd, 107.12 mph.

REVOLUTIONS PER MINUTE

Now let's suppose you want to know the rpm at a given mph. For that, the formula would be:

$$RPM = \frac{mph \times gear\ ratio \times 336}{tire\ diameter}$$

What would the Mustang's rpm be out on a rural highway, running in 5th gear at the legal max of 65 mph?

$$RPM = \frac{65 \times 0.68 \times 3.08 \times 336}{26}$$

$$RPM = \frac{45741.696}{26}$$

The answer is 1759 rpm. The engine is loafing when cruising in overdrive, and that means good fuel economy.

How do changes in gearing or tire size affect the engine or vehicle speed of a strong runner like the Pontiac Firebird? You can find out by applying the formulas in the text.

OVERALL GEAR RATIO

But suppose you'd rather gear the car for maximum response at highway speed. At 65 mph, you want to be able to downshift from 5th to 4th and be at the Mustang V-8's torque peak, which happens to be 3000 rpm, in order to have optimum passing ability. The overall gear ratio, which in direct drive 4th gear would simply be the final drive ratio, is now the unknown and the formula is:

$$\text{Gear Ratio} = \frac{\text{rpm} \times \text{tire diameter}}{\text{mph} \times 336}$$

The figures for the Mustang would be:

$$\text{Gear Ratio} = \frac{3000 \times 26}{65 \times 336}$$

$$\text{Gear Ratio} = \frac{78000}{21840}$$

You should have a 3.57 final drive ratio in order to reach 3000 rpm at 65 mph in 4th gear. Final drive gears don't come in an infinite variety of ratios, though, and the closest you're likely to come to the ideal figure would be 3.54. To see what the rpm would be with that ratio, let's go back to the formula for rpm:

$$\text{RPM} = \frac{65 \times 3.54 \times 336}{26}$$

$$\text{RPM} = \frac{77313.6}{26}$$

That gives you 2973.6 rpm, which is certainly close enough for all practical purposes.

TIRE DIAMETER

Finally, there's the formula for tire diameter:

$$\text{Tire Diameter} = \frac{\text{mph} \times \text{gear ratio} \times 336}{\text{rpm}}$$

A circular calculator or "dream wheel" for finding final drive ratio when rpm, mph and tire size are all known is available from Mark Williams Enterprises, a prominent supplier of high-performance axles and gears in Louisville, Colorado. For pricing and ordering information, call (303) 665-6901.

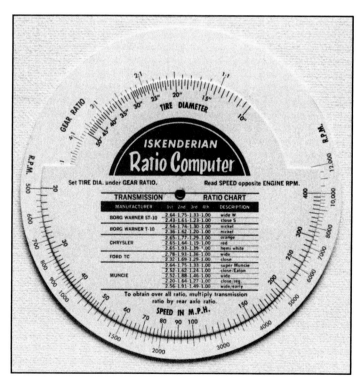

A similar "dream wheel" for finding final drive ratio that also includes a separate dial on the flip side for finding engine displacement when bore and stroke are known is offered by Iskenderian Racing Cams, a famous camshaft grinder in Gardena, California. Phone (213) 770-0930 for details.

You're not likely to use this one as often as those for mph, rpm and gear ratio, but let's try an example anyway.

Going back to modifying the Mustang to turn 3000 rpm at 65 mph in 4th gear, you could achieve a similar effect by changing the tire size instead of the final drive ratio:

$$\text{Tire Diameter} = \frac{65 \times 3.08 \times 336}{3000}$$

$$\text{Tire Diameter} = \frac{67267.2}{3000}$$

The Lowdown—According to the formula, you'd need tires with a diameter of 22.4 inches to do the job. That would convert the Mustang into something of a low rider, which might have ground clearance problems on steep driveways. But, as a quick and dirty way of achieving the wanted relationship between mph and rpm, it would work.

There is another problem, though. With the change in tire size, the speedometer will no longer read accurately. You can find what the error is with the techniques described in Chapter 12. Or you can figure out what it should read with a formula presented in Chapter 14.

Meanwhile, you'll find the formulas that have been discussed here—especially those for mph, rpm and gear ratio—among the most useful you can know, and I bet you'll be turning to them often. When you get used to working with them regularly, you'll probably wonder how you ever got along without them.

Table 13

FORMULAS FOR MPH, RPM, GEARS & TIRES

$$mph = \frac{rpm \times tire\ diameter}{gear\ ratio \times 336}$$

$$rpm = \frac{mph \times gear\ ratio \times 336}{tire\ diameter}$$

$$gear\ ratio = \frac{rpm \times tire\ diameter}{mph \times 336}$$

$$tire\ diameter = \frac{mph \times gear\ ratio \times 336}{rpm}$$

Tire Sizes & Their Effects 14

The two-wheel-drive Toyota XtraCab SR5 pickup has low-profile tires that reduce the truck's center of gravity for good handling and stability on the street and highway.

Suppose you're planning to replace the tires on your car or truck with bigger ones. Say, for example, you have a high-performance car and need bigger rubber at the rear for better traction at the drag strip. Or you have a four-wheel-drive vehicle and want to increase its ground clearance for off-road driving.

Whatever your reason for wanting bigger tires, there are a couple of important questions to consider.

First, what effect will the new tires have on the overall gearing? The vehicle may not respond the way you're used to, and you may find it necessary to downshift more frequently. You may even have to change the final-drive ratio in order to retain—or regain—the original level of performance.

Second, the speedometer will read too slow. On the highway, you may be deceived into thinking you're cruising at the legal maximum, when actually you're well over it and ripe for a speeding ticket.

95

The four-wheel-drive XtraCab SR5 has larger diameter tires that provide obvious benefits in ground clearance for rough, off-highway terrain. But the bigger tires also require different gearing and recalibration of the speedometer and odometer.

TIRE DIAMETER

Fortunately, if you know the diameters of both the new and old tires and the vehicle's existing final-drive ratio, you can calculate the effects the bigger tires will have ahead of time. Your local tire dealer should have charts showing the diameters of the various tires he carries. Or you can simply apply a tape measure to one of the tires currently on your car or truck, and to one of those you're considering as replacements.

Section Height & Width—You may also be able to figure out the diameters of the tires from their respective sizes. In the old days when 6.00x16 was the standard size on many popular cars, that was easy. The tire's *section height*—the distance between the edge of the rim and the face of the tread—and the *section width*—the distance between the sidewalls on either side—were about the same.

In the case of a 6.00x16, the section height and width were both about 6.0 inches and the tire was mounted on a 16-inch wheel. To find the diameter, you simply multiplied the section height, 6.0 inches, by 2, giving you 12 inches, and added that to the wheel rim diameter, 16 inches, for an overall figure of 28 inches.

For modern heavy-duty truck tires, the sizing system is even more straightforward. For example, a 31x11.50x15 tire will have a nominal diameter of 31 inches and width of 11.5 inches, and will fit a 15-inch wheel. Frankly, though, I'm hedging with that word "nominal," because tire industry standards allow up to a 7.0 percent variation from specified dimensions.

Aspect Ratio—Modern passenger car tires and light-duty truck tires aren't sized quite so simply. In most cases, their section height and width are no longer alike. The height is usually much less than the width, and the relationship between the two—the *aspect ratio*—is an important part of their specs. The aspect ratio is the percentage the section height is of the section width.

Generally speaking, passenger car and light truck tires are now produced in metric sizes that indicate the section width in millimeters, the aspect ratio in percent and the wheel rim diameter in inches.

Example—As a case in point, let's take an LT235/75R15 tire. The "LT" means it's a light truck tire; if it were a passenger car unit, it would have the initial "P" instead. Similarly, the "R" means it's a radial, while a "B" would indicate it's bias belted. The "235" is the section width in millimeters and the "75" is the aspect ratio, indicating that the the section height is 75 percent of the section width. Finally, the "15" is the rim diameter in inches.

METRIC TIRE DIAMETER

To find the diameter in inches of a metric size tire, you must first find the section height in inches. To do that, you convert the section width, 235 millimeters in our example, to inches by dividing it by 25.4, the number of millimeters in an inch. Then you convert the aspect ratio, 75, to a decimal figure by dividing it by 100.

Multiply the quotients of these two calculations together to find the section height in inches. Double that figure and add the wheel rim diameter, which is already given in inches, and the result will be the diameter of the tire in inches.

Expressed as a formula, that would be:

$$\text{Tire Diameter} = 2 \times \frac{\text{section width}}{25.4} \times \frac{\text{aspect ratio}}{100} + \text{rim dia.}$$

That can be simplified somewhat to:

$$\text{Tire Diameter} = 2 \times \frac{\text{section width} \times \text{aspect ratio}}{2540} + \text{rim dia.}$$

Plug in the appropriate specs for a LT235/75R15 tire:

$$\text{Tire Diameter} = 2 \times \frac{235 \times 75}{2540} + 15$$

$$\text{Tire Diameter} = 2 \times 6.9 + 15$$

That would work out to a section height of 6.9 inches and an overall diameter of 28.88 inches which, of course, could be rounded up to 28.9 inches.

Checking the specs for 3 different BFGoodrich radial light truck tires available in the LT235/75R15 size—the Sport Truck T/A, All Terrain T/A and Mud-Terrain T/A—you would find their diameters are listed as 28.9, 28.98 and 29.09 inches, respectively, all of them within less than 1.0 percent of our calculated figure for that size.

I have assumed the wheel rim meets a tire industry standard requiring a rim width that is 70 percent of the tire section width. For every 0.5 inch increase or decrease in rim width, there would be a corresponding 0.2 inch increase or decrease in the section width of the mounted tire.

EFFECTIVE DRIVE RATIO

To find what the effective overall drive ratio would be with a given increase in tire diameter, the formula is:

$$\text{Effective Ratio} = \frac{\text{old tire diameter}}{\text{new tire diameter}} \times \text{original ratio}$$

Off-road tires come in a wide variety of sizes, as is evident from this selection offered by Dick Cepek, Inc. (photo courtesy Spencer Murray).

Example—Suppose you have a set of 28.9-inch LT235/75R15s on a four-wheel-drive truck with a 3.08 final-drive ratio and, to increase the ground clearance, you want to replace them with 33-inch 33x12.50x15s. To find the effective drive ratio with the bigger tires, the figures would be:

$$\text{Effective Ratio} = \frac{28.9}{33} \times 3.08$$

$$\text{Effective Ratio} = 0.8757576 \times 3.08$$

With the bigger tires, the effective ratio is only 2.70! That's enough of a change to cause a noticeable loss in responsiveness.

EQUIVALENT DRIVE RATIO

To find the final-drive ratio needed with the new tires to provide the equivalent of the vehicle's performance with the original tires, the formula is:

$$\text{Equivalent Ratio} = \frac{\text{new tire diameter}}{\text{old tire diameter}} \times \text{original ratio}$$

Note that the positions of the tire diameters in this formula are reversed from their positions in the formula for effective ratio. In the case of the switch from 28.9- to 33-inch tires on a vehicle with a 3.08 final drive, the figures would be:

$$\text{Equivalent Ratio} = \frac{33}{28.9} \times 3.08$$

$$\text{Equivalent Ratio} = 1.1418685 \times 3.08$$

That works out to 3.51, so a set of final-drive gears in the 3.50-plus range would be needed to restore the lost responsiveness.

SPEEDOMETER CORRECTION

Assuming the speedometer was accurate with the original tires, the formula for determining what the actual speed is at any indicated speed with the bigger tires is:

$$\text{Actual MPH} = \frac{\text{new tire diameter}}{\text{old tire diameter}} \times \text{indicated mph}$$

Following the swap from 28.9-inch to 33-inch tires, what would the actual speed be at an indicated 65 mph?

$$\text{Actual MPH} = \frac{33}{28.9} \times 65$$

$$\text{Actual MPH} = 1.1418685 \times 65$$

The actual speed, rounded down, would be 74.2 mph. In many parts of the country, that's more than enough to attract the attention of the state highway patrol. Of course,

The high ground clearance afforded by oversize tires helps a four-wheel-drive vehicle like this Dodge Ramcharger to wend its way over irregular terrain.

High clearance tires are essential for coping with rocky off-highway conditions without damaging the vehicle's undercarriage components, as the driver of this Ramcharger is attempting to do.

you can get a new speedometer drive gear, as suggested in Chapter 12. In the meantime, though, it would be helpful to know what the indicated speed would be at an actual 65 mph, using the formula:

$$\text{Indicated MPH} = \frac{\text{old tire diameter}}{\text{new tire diameter}} \times \text{actual mph}$$

Or, using the figures in the ongoing example:

$$\text{Indicated MPH} = \frac{28.9}{33} \times 65$$

$$\text{Indicated MPH} = 0.8757576 \times 65$$

At an actual 65 mph, the indicated speed would be 56.9 mph. If you keep the speedometer reading under that figure, you won't get stopped for breaking the 65-mph limit.

DOWNSIZE TIRES

Although most auto enthusiasts would be more likely to want an increase in tire size than a decrease, the formulas for effective and equivalent drive ratios and for actual and indicated mph would be equally valid for a change to smaller tires.

As an example, consider that improbable Mustang low rider described in Chapter 13. To recall its pertinent specs, its old tires were 26 inches in diameter and its new ones 22.4 inches, while its final drive ratio was 3.08. To find its effective drive ratio with the smaller tires, the equation would be:

$$\text{Effective Ratio} = \frac{26}{22.4} \times 3.08$$

$$\text{Effective Ratio} = 1.1607143 \times 3.08$$

And the effective ratio would be 3.575. To find the equivalent ratio:

$$\text{Equivalent Ratio} = \frac{22.4}{26} \times 3.08$$

$$\text{Equivalent Ratio} = 0.8615385 \times 3.08$$

The theoretical ratio would be 2.6535385 or, rounded down 2.65, and the closest gearset to that in the Mustang parts catalog is a 2.73.

With an uncorrected speedometer, what would the actual speed be at an indicated 65 mph?

$$\text{Actual MPH} = \frac{22.4}{26} \times 65$$

$$\text{Actual MPH} = 0.8615385 \times 65$$

The answer is an even 56 mph. Finally, what would the indicated speed be at an actual 65 mph?

$$\text{Indicated MPH} = \frac{26}{22.4} \times 65$$

$$\text{Indicated MPH} = 1.1067143 \times 65$$

And the answer this time is 75.446429 or, rounded up, 75.45 mph.

Note that earlier, with bigger tires, the speedometer was slow. Now, with smaller ones, the instrument is fast.

Why the Diameter?—Mathematically, the relationship between final-drive ratio or speedometer reading with different tires is in direct proportion to the differences in the sizes of the tires. The formulas would work equally well with the respective rolling radii or circumferences. You could also use the respective revolutions per mile of the tires; in that case, though, the results would be in inverse rather than direct proportion.

But, generally, the tires' diameters are among the easiest specs to find, even with the mumbo jumbo needed to calculate the diameter with a metric size, and that's why they've been used here.

Table 14

FORMULAS FOR TIRE SIZES & THEIR EFFECTS

$$\text{tire diameter} = 2 \times \frac{\text{section width} \times \text{aspect ratio}}{2540} + \text{rim diameter}$$

$$\text{effective ratio} = \frac{\text{old tire diameter}}{\text{new tire diameter}} \times \text{original ratio}$$

$$\text{equivalent ratio} = \frac{\text{new tire diameter}}{\text{old tire diameter}} \times \text{original ratio}$$

$$\text{actual mph} = \frac{\text{new tire diameter}}{\text{old tire diameter}} \times \text{indicated mph}$$

$$\text{indicated mph} = \frac{\text{old tire diameter}}{\text{new tire diameter}} \times \text{actual mph}$$

Average MPH & MPG **15**

Most drivers are already familiar with the formulas for miles per gallon and miles per hour. Knowing how to manipulate them properly can add to both the efficiency and enjoyment of highway travel.

The formulas for average fuel mileage and average highway speed are familiar to most drivers. In fact, the expressions used to describe these two quantities—miles per gallon and miles per hour—are statements of their respective equations.

Miles per gallon means the distance driven in miles divided by the amount of fuel used in gallons, or:

$$\text{Miles Per Gallon} = \frac{\text{miles}}{\text{gallons}}$$

Miles per hour means the distance driven in miles divided by the time of the trip in hours, or:

$$\text{Miles Per Hour} = \frac{\text{miles}}{\text{hours}}$$

Now let's take a closer look at average fuel mileage.

MILES PER GALLON

Suppose you drive from Boston to Washington, D.C. Your first stop is in New York City, which is 208 miles from Boston, and your car uses 10.4 gallons of fuel. What would the average fuel mileage be?

$$\text{Miles Per Gallon} = \frac{208}{10.4}$$

You got an even 20 miles per gallon.

Fuel Range—Suppose your car has an 18-gallon tank. If you hadn't refilled in the Big Apple, how far could you have gone before you had to stop for fuel? In other words, what would your fuel range be in miles?

$$\text{Miles} = \text{miles per gallon} \times \text{number of gallons}$$

Which, in this case, would be:

$$\text{Miles} = 20 \times 18$$

Or 360 miles. You've already driven 208 miles, so you could've gone up to 152 miles farther before running out of fuel. You would've probably had to start looking for a filling station that would accept one of your credit cards when you passed through Philadelphia, which is 106 miles from New York. You wouldn't have made it as far as Baltimore, which is another 96 miles, or 202 miles from New York. That's 50 miles beyond your range.

Predicting Fuel Consumption—Washington is 239 miles from New York. At 20 miles per gallon, how much fuel will the car use for the trip? To find out, divide the miles by the miles per gallon:

$$\text{Gallons} = \frac{\text{miles}}{\text{miles per gallon}}$$

Or, in this case:

$$\text{Gallons} = \frac{239}{20}$$

Which would be 11.95 gallons. You've already used 10.4 gallons getting from Boston, so your overall fuel consumption would be 22.35 gallons for the total distance of 447 miles from Boston to Washington.

MILES PER HOUR

On the stretch from Boston to New York, you took 4-1/4 hours to cover the 208 miles. To find your average speed, you convert that 4-1/4 to 4.25 and plug it into the formula for average miles per hour:

$$\text{Miles Per Hour} = \frac{208}{4.25}$$

On an 8-digit calculator, that would be 48.941176, which can be rounded up to 49 mph.

What if your time was 4 hours and 35 minutes, or 4:35? Can you run that as:

$$\text{Miles Per Hour} = \frac{208}{4:35}$$

No you can't. That 4:35 is 4 and 35/60, not 4 and 35/100. However, with a scientific calculator, you can run it this way:

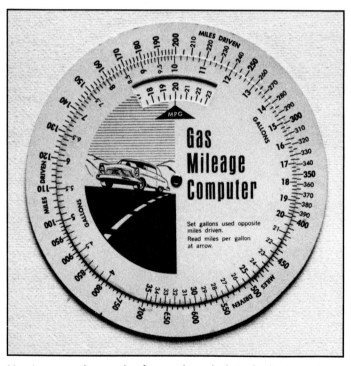

Here's a typical example of a circular calculator for finding miles per gallon. Devices like this have long been used as giveaway items by automakers and aftermarket equipment and accessory manufacturers.

$$\text{Miles Per Hour} = \frac{208}{(4 + 35/60)}$$

That would be the equivalent of:

$$\text{Miles Per Hour} = \frac{208}{4.5833333}$$

Which would work out to 45.381818 or, rounded down, 45.38 mph.

Distance & Time—Suppose you want to know how far you've gone If you drive at a given speed for a given number of hours. Multiply the miles per hour by the hours:

$$\text{Miles} = \text{miles per hour} \times \text{hours}$$

If you drive at an average of 45 mph for 2-1/2 hours, how far would you go?

$$\text{Miles} = 45 \times 2.5$$

You would cover 112.5 miles. If you drive at a given speed for a given number of miles, how many hours will it take? To find out, divide the miles by the miles per hour:

Iskenderian's Average Speed Computer is another "dream wheel," this one for calculating lap speeds. On the other side is a dial for figuring compression ratio. Call (213) 770-0930 for pricing and ordering information.

$$\text{Hours} = \frac{\text{miles}}{\text{miles per hour}}$$

As an example, suppose you continue the 106 miles from New York to Philadelphia at an average of 45 miles per hour:

$$\text{Miles} = \frac{106}{45}$$

That works out to 2.3555556, which can be rounded up to 2.36 hours.

Hundredths to Minutes—That 2.36 hours is 2-36/100 hours, not 2 hours and 35 minutes. To convert hundredths of an hour to minutes, or sixtieths of an hour, i.e., to convert from a centestimal to a sexagesimal fraction, you can set up a simple algebraic equation, using the letter Y as the unknown, to find how many sixtieths the hundredths would equal:

$$\frac{36}{100} = \frac{Y}{60}$$

That would be 100 x Y = 36 x 60 or 100 x Y = 2160 or:

$$Y = \frac{2160}{100}$$

And Y works out to be 21.6 minutes. That 0.6 of a minute is, of course, 6/10 of a minute, not 6.0 seconds. But by now, you know how to convert that 6/10 of a minute to seconds. Don't you?

RACEWAY LAP TIMES & AVERAGE SPEEDS

A similar set of formulas can be used to calculate performance on an oval track or road course. Average lap speed in miles per hour can be found by multiplying the lap distance in miles by 3600—the number of seconds in an hour—and dividing the result by the lap time in seconds:

$$\text{Miles Per Hour} = \frac{\text{miles} \times 3600}{\text{seconds}}$$

On a race course, average speed in miles per hour can be calculated from lap time in seconds and course length in miles. Yes, this is the Chevrolet Lumina driven by Tom Cruise as Cole Trickle in the motion picture Days of Thunder.

Suppose you're running at Indianapolis and averaging 40 seconds a lap around the 2.5-mile track. What's your average speed?

$$\text{Miles Per Hour} = \frac{2.5 \times 3600}{40}$$

$$\text{Miles Per Hour} = \frac{9000}{40}$$

You're cooking at an even 225 mph. What kind of a lap time would you have to clock in order to average 230 mph? To find that, again multiply the lap distance by 3600 but,

this time, divide by the speed you want to reach:

$$\text{Seconds} = \frac{\text{miles} \times 3600}{\text{miles per hour}}$$

To find the lap time needed for 230 mph at Indy:

$$\text{Seconds} = \frac{2.5 \times 3600}{230} = \frac{9000}{230}$$

You must get around the track in 39.130435 seconds—pretty brisk for the brickyard!

Table 15

FORMULAS FOR AVERAGE MPH & MPG

$$\text{miles per gallon} = \frac{\text{miles}}{\text{gallons}}$$

$$\text{miles} = \text{miles per gallon} \times \text{gallons}$$

$$\text{gallons} = \frac{\text{miles}}{\text{miles per gallon}}$$

$$\text{miles per hour} = \frac{\text{miles}}{\text{hours}}$$

$$\text{miles} = \text{miles per hour} \times \text{hours}$$

$$\text{hours} = \frac{\text{miles}}{\text{miles per hour}}$$

$$\text{raceway miles per hour} = \frac{\text{miles} \times 3600}{\text{seconds}}$$

$$\text{seconds} = \frac{\text{miles} \times 3600}{\text{miles per hour}}$$

Blood Alcohol Concentration 16

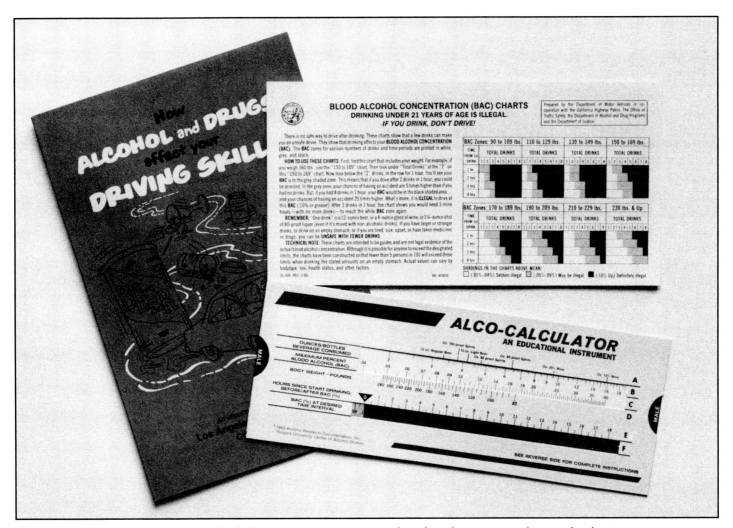

The formula for estimating blood alcohol concentration presented in this chapter provides results that are consistent with most methods for calculating BAC in booklets and charts published by motor vehicle departments, insurance companies and other authorities.

Drunk driving is one of our nation's most frequent causes of death, especially among younger drivers. According to a report released by the National Highway Traffic Safety Administration (NHTSA) in late 1987:

"An alcohol-related fatality occurs every 22 minutes in this country, or 66 a day. Hundreds more are injured. Last year [1986], 46,056 people died in traffic crashes and 52 percent of those fatalities involved alcohol. . . .

"Drunk driving remains the number one killer on our roads and of our young people between the ages of 5 and 34."

Those are frightening statistics and the sad truth is that they're unnecessary. If people understood how few drinks it takes to impair their ability to operate a vehicle and would not get behind the wheel when they've been imbibing, thousands of highway accidents could be avoided and thousands of lives saved.

FOUR FACTORS

When a person has been drinking an alcoholic beverage, the measure of how much alcohol he or she has consumed is *blood alcohol concentration* or BAC, which is the percentage of alcohol in the bloodstream. With the proper testing facilities, BAC can be determined directly from the amount of alcohol in the blood, or it can be correlated from the presence of alcohol in the breath or urine.

It can also be estimated mathematically from four factors: 1) the amount in ounces the person has had to drink; 2) the percentage of alcohol in the particular beverage the person has been drinking; 3) the person's body weight; and 4) the time in hours the person has been drinking.

Limitations—The formula isn't precise, because it's based on averages and doesn't consider such factors as individual differences in metabolism. Furthermore, it doesn't take into account some variables which could affect the amount of alcohol absorbed into the bloodstream but which can't be measured readily, such as how much food a person might've eaten before or while drinking.

So figures derived from the formula wouldn't stand up in court against evidence from blood, breath or urine tests. However, they would be consistent with most charts for calculating BAC published by motor vehicle departments, insurance companies and other authorities. And, certainly, the formula has educational value for showing the comparative effects of different amounts of different alcoholic beverages.

Alcohol in the Bloodstream—The volume of blood in a human body corresponds to between 7 and 8 percent—or an average of 7.5 percent—of the body weight. Thus, a larger and therefore heavier person can absorb more alcohol than a smaller, lighter one before reaching a specific BAC. (However, this doesn't apply to someone who's overweight. Such a person's blood supply doesn't increase in proportion to the excess weight.)

LEGAL LIMITS

In most of the United States, a person is considered intoxicated if his or her BAC is 0.10 or more and is subject to arrest. That figure of 0.10 means that 1/10 of 1.0 percent of the bloodstream is alcohol or, to rephrase it, there is 1 part alcohol in 1,000 parts of blood. According to the California Department of Motor Vehicles, a person with a BAC of 0.10 or higher is 25 times more likely than a sober person to have an accident!

At this writing, there are four states—California, Oregon, Utah and Maine—where a person is considered intoxicated and subject to arrest with a BAC of only 0.08 or 8/100 of 1.0 percent. Several other states are considering reducing the legal limit from 0.10 to 0.08.

In areas of commercial transportation under federal authority, such as interstate trucking, flying and boating, a BAC of only 0.04 is considered the threshold of intoxication.

There are other countries which have even stricter laws. In Sweden, for example, a driver can be arrested if he or she has a BAC of as little as 0.02!

Gray Area—In most of the United States, a BAC of less than 0.04 isn't considered hazardous, but the range from 0.04 to 0.08 is controversial. It's a gray area and, in most states, a driver whose BAC is within those limits isn't subject to arrest. However, such a driver is considered impaired and, again citing the California DMV, his or her chances of having an accident are 5 times greater than normal. Obviously, someone with a BAC between 0.04 and 0.08 would be wiser not to drive, even though he or she may be legally free to do so.

Irreversible Effects—Once alcohol enters the bloodstream, there's no way to reverse its effects. The drinker simply has to wait until the alcohol disappears from his or her system. The rate at which that occurs can vary widely, from 0.007 to 0.04 percent of the BAC per hour. The average is about 0.015 percent per hour.

Percentage of Alcohol—Finally, there's the volume of alcohol in the beverage. For distilled spirits, such as whiskey and vodka, the percentage of alcohol by volume is shown on the label. Another measure of the strength of distilled spirits on the label is the proof, which is simply twice the percentage of alcohol. Most spirits range from 40 to 50 percent alcohol, or from 80 to 100 proof.

Wine labels also state the percentage of alcohol by volume. For dry wines, the figure is usually about 12.5 percent and, for sweet wines, 20 percent.

Beer is a problem because, unlike distilleries and wineries, breweries are not required to indicate the alcoholic content of their products. Further complicating matters is the fact that the legally allowed amount of alcohol in beer varies in many states, and that some states

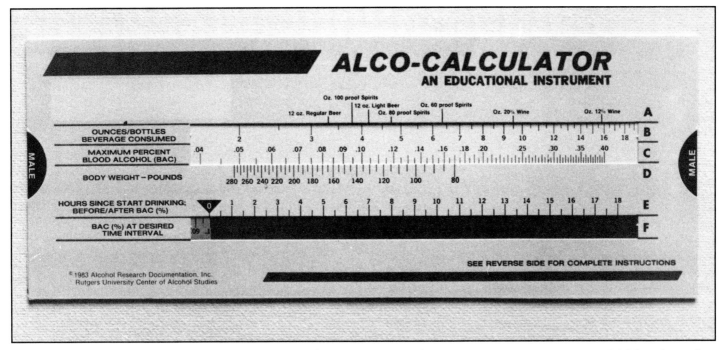

The Alco-Calculator is a slide rule developed by the late Professor Leon A. Greenburg at Rutgers University to estimate BAC from basically the same factors as the formula in the text. For more information about ordering the Alco-Calculator, write Publications Division, Rutgers Center of Alcohol Studies, P.O. Box 969, Piscataway, New Jersey 08854.

specify a limit in percentage by weight rather than by volume.

In beer, the percentage of alcohol by weight is 80 percent, or 4/5, of the percentage by volume. Thus, a brand that is 3.2 percent alcohol by weight would be 4.0 percent by volume. That's a fairly common limit, so 4.0 percent is used here as the norm for beer.

BAC FORMULA

To find BAC, multiply the amount consumed in ounces by the percent of alcohol by 0.075 (representing the volume of blood in the body), divide by the body weight, then subtract the number of hours the person has been drinking multiplied by 0.015 (the percentage BAC drops per hour). To put all that in the form of an equation:

$$BAC = \frac{(ounces \times \% \ alcohol \times 0.075)}{weight} - (hours \times 0.015)$$

Example—To see how that works, take as an example a 180-pound man who consumes a 6-pack of 12-ounce beers, or a total of 72 ounces, with a 4.0-percent alcoholic content in 2 hours. The figures would be:

$$BAC = \frac{(72 \times 4 \times 0.075)}{180} - (2 \times 0.015)$$

$$BAC = \frac{21.6}{180} - 0.03$$

$$BAC = 0.12 - 0.03$$

That's a blood alcohol concentration of 0.09. In states where a BAC of 0.09 is considered evidence of intoxication, a 180-pound man who downs a 6-pack of beers with a 4.0 percent alcohol content in 2 hours and then

111

gets behind the wheel of his car can be arrested for driving under the influence.

Supposed he'd been more temperate and drank only 2 beers, or 24 ounces, in those 2 hours:

$$BAC = \frac{(24 \times 4 \times 0.075)}{180} - (2 \times 0.015)$$

$$BAC = \frac{7.2}{180} - 0.03$$

$$BAC = 0.04 - 0.03$$

Under these conditions, his BAC would be only 0.01, 1/10 of the definition of intoxication in most states and 1/8 of it in the stricter states, so he should be able to drive safely.

One criticism sometimes made of this formula is that those who need it the most aren't likely to be in condition to work with it! That's not really the point. It's not meant to be used during the evening happy hour at the local tavern. Rather, it's intended as an educational tool.

Try running different combinations of the four factors. You'll discover that equivalent drinks—12 ounces of 4.0-percent beer, 4.0 ounces of 12.5-percent wine, or 1-1/4 ounces of 40-percent whiskey—produce similar results. It isn't what you drink that's important; it's how much. You can get just as intoxicated on beer or wine as you can on distilled spirits.

Most important of all, if you're a drinker yourself, or you know someone who is, calculate your, or their, BAC for a typical night out on the town. You're likely to find the results—uh—sobering!

TABLE 16

FORMULAS FOR BLOOD ALCOHOL CONCENTRATION

$$BAC = \frac{(\text{ounces} \times \% \text{ alcohol} \times 0.075)}{\text{weight}} - (\text{hours} \times 0.015)$$

Auto Math on Your Computer 17

```
100 REM disk file BORESTRO
105 CLS
110 PRINT TAB(6);"Increases in Bore or Stroke"
115 PRINT
120 INPUT"Present bore";B1
125 INPUT"Present stroke";S1
130 INPUT"Number of cylinders";C
135 INPUT"Cubic inch limit";D2
140 P=.7853982
145 D1=P*B1^2*S1*C
150 B2=SQR(D2/(P*S1*C))
155 S2=D2/(P*B1^2*C)
160 B3=B2-B1
165 S3=S2-S1
170 PRINT
175 PRINT"Engine's present displacement"
180 PRINT"is";INT(D1*1000+.5)/1000;"cubic inches."
185 PRINT
190 PRINT"Within a";D2;"cubic inch limit:"
195 PRINT
200 PRINT"Bore may be increased up to";INT(B3*1000+.5)/1000;"inch"
205 PRINT"from";B1;"to"INT(B2*1000+.5)/1000;"inches, or"
210 PRINT
215 PRINT"Stroke may be increased up to";INT(S3*1000+.5)/1000;"inch"
220 PRINT"from";S1;"to";INT(S2*1000+.5)/1000;"inches."
225 PRINT
230 END
```

Fig. 17a. This is a printout of a working BASIC program for determining possible increases in bore or stroke within a given displacement limit. If you enter it on your computer, type it exactly as shown. The quote marks, colons, semi-colons, asterisks and so forth are all where they are for specific programming reasons. If you attempt to change or reposition them, you will probably render the program inoperative.

If you're computer literate as well as a car enthusiast—a hacker as well as a hot rodder—you might like to try writing BASIC computer programs using some of the automotive mathematics presented in this book.

BASIC, as you may already know, stands for Beginner's All-purpose Symbolic Instruction Code. It's the most popular computer programming language. There isn't the space here to discuss BASIC in detail, and I'll assume you're familiar with the version of it for your particular computer. If not, I urge you to get a good textbook on the subject.

GEE WHIZ

The form of the language used on the most brands of computers is Microsoft GW-BASIC. The GW, believe it or not, stands for Gee Whiz. GW-BASIC comes packaged with MS-DOS, or Microsoft Disk Operating System, which is used on so-called IBM clones and is the key to a computer's compatibility with IBM software.

The IBM itself uses a variation of GW-BASIC called BASICA—for Advanced version. For all practical purposes, GW-BASIC and BASICA are the same, and 99.9 percent of the programming written with one of them will work with the other.

Consequently, the programming presented here is written in GW-BASIC. I've tried to keep it simple, though, to facilitate translating it to other forms of the language. And, toward the end of the chapter, I've included a few notes for adapting the program to the version of BASIC used on another popular computer, the Commodore 64.

CHOOSING MATH FOR PROGRAMMING

Every formula in this book can be adapted to BASIC, but computerizing the simpler equations can be overkill. For example, there are several books on BASIC that contain programs for computing a car's miles per gallon. The authors of such works are trying to demonstrate the computer language with programming about everyday concerns and, certainly, most of us do care about how many miles per gallon our cars get.

But, as you've seen in Chapter 15, figuring miles per gallon involves the simplest arithmetic, i.e., dividing the number of miles driven by the number of gallons consumed. You can do that on a calculator or even in your head in less time than it would take to load a computer program. Such programming complicates something that's simple, when what a computer should be used for is simplifying things that are complicated.

Suppose you use a particular formula repeatedly, or a formula that's especially complex, or even simple formulas in elaborate combinations. In any of those cases, the computer can be used to simplify the complications.

DEVELOPING A PROGRAM

As a case in point, consider the formulas for finding an engine's displacement, stroke and bore, as shown in Table 1 on page 7.

I pointed out in Chapter 1 that these equations are particularly useful for figuring out how much the bore or stroke can be increased while staying within a given displacement limit, such as the maximum for a specific racing class. But doing that on a calculator can be tedious, and it's easy to lose track of your figuring. So this is an excellent subject for a computer program.

In all three of these formulas, *pi*/4, or 0.7853982, is the only constant. The displacement, bore, stroke and number of cylinders are all variables, i.e., they will differ from one application to another.

In GW-BASIC, you can use variable names up to 40 characters long, so you can spell them out in plain English in programs, just as I've done with most of the variables in the formulas throughout this book.

However, you can also abbreviate them to as few as one or two characters, and that will save a lot of typing when you first enter a program on the computer. It will also make it easier to adapt the program to other forms of BASIC, some of which can accommodate variable names of only one or two characters.

When you do use abbreviated variable names, it's good programming practice to make them the initial letters of the individual factors—D for displacement, B for bore, S for stroke and C for cylinders. You'll find that done in simplified versions of the formulas for displacement, bore and stroke shown in Table 17, page 120.

Getting Down to BASIC—Now let's convert the formulas to BASIC.

You can spare yourself the tedium of repeatedly typing in the one constant by defining the value of, say, P as *pi*/4, or 0.7853982, and then using P in the program wherever *pi*/4 appears in the formulas.

In BASIC, you use an asterisk (*) as a multiplication sign instead of an x. To square a variable, you type an upright arrow of carat (^) after it and then the number 2. (If you want to cube a variable, type a 3 after the carat.) To find a square root, type the BASIC function SQR and then, in parentheses, the value from which you want the root.

In Fig. 17b, you'll find the three formulas rewritten with these changes.

Fig. 17a shows the program itself. In lines 145, 150 and 155, you'll see variations of the formulas in Fig. 17b, with

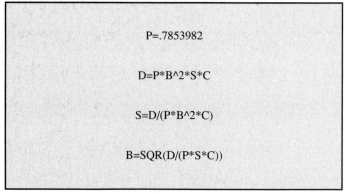

P=.7853982

D=P*B^2*S*C

S=D/(P*B^2*C)

B=SQR(D/(P*S*C))

Fig. 17b. These are the formulas for displacement, stroke and bore, as well as the constant for *pi*/4 (P), as they appear in BASIC.

one notable difference. You'll be dealing with the measurements of an engine in both present and modified form, so you should use numbered versions of the variable names—D1, B1 and S1 for the present specs, and D2, B2 and S2 for their modified equivalents. You can leave C as is, though, because you're not likely to be altering the number of cylinders!

How It Works—Let's see how the program works before analyzing why it works, using the same example as in the original explanation of the formulas in Chapter 1.

You have a Ford 351 V-8 which, with its 4.0-inch bore and 3.5-inch stroke, actually displaces closer to 352 cubic inches than 351. You want to rework the engine for a racing class that has a displacement limit of 366 cubic inches, and you want to know how much you can increase the bore with the stock stroke, or how much you stretch the stroke with the stock bore. When you run the program and enter the known data, the computer video monitor will display the results as shown in Fig. 17c.

Suppose you've already bored the cylinders 0.060-inch over and you want to know how much room you have left for increasing the stroke. Run the program again and, in

Fig. 17c. Here is an actual run of the program using the specs of the Ford 352, uh, 351 cubic-inch V-8. As the results show, that engine could be bored up to 0.08 inch or stroked up to 0.141 inch while staying within a limit of 366 cubic inches.

response to "Present bore?" enter 4.06, then enter the rest of the data as before. As you can see from the display shown in Fig. 17d, you won't have much of a margin for further modifications.

In either case, you've had to type in only four numbers. Stop and think about how much poking you'd have to do on a calculator to find the same information!

Further, on a computer, all the figuring happens almost instantly. Even with the time you'd need to load the program from a disk into the computer, you'd have your results sooner than you would with a calculator.

Why It Works—Now go back to Fig. 17a and take a closer look at the program. One of the first things that may strike you is the extensive use of the word PRINT. When BASIC was first developed in 1964, computer video monitors were still a rarity, and program listings and results were displayed by actually printing them out on a teletypewriter.

Today, PRINT is an instruction to show on the monitor whatever follows the word PRINT on the program line. If nothing comes after it, it provides a blank line in the display. That feature can be used to spread out a display either on a monitor or a printout for easier readability.

In a mathematical computer program, there are operations that have to be performed in proper sequence. For example, you obviously can't find displacement until you've entered bore, stroke and number of cylinders, and processed them with the appropriate formulas. That's what those numbers at the beginnings of the program lines are about. They tell the computer the order in which the lines are to be executed.

The program begins at line 100 with a REM, for *remark*, to identify the name used to file it on disk. In GW-BASIC, the file name can be no more than 8 characters, which makes it difficult to write a very descriptive name. The REM line is non-functioning and, when the program is run, the name doesn't appear as part of the display.

CLS in line 105 clears the screen and moves the cursor to the home position at top left.

PRINT TAB(6) in line 110 is a tabulating function that starts the headline "Increases in Bore or Stroke" 6 spaces in from the left edge of the screen.

Line 115 provides a blank line between the headline and the first working program line.

In lines 120 through 165, you enter and process the engine specs. As you've already seen, the INPUTs in lines 120 through 135 ask for the known factors. The program pauses at each of these lines until a number is entered and the enter or return key is pressed. When you've completed the entries and hit the enter or return key for the fourth and final time, the program goes to work.

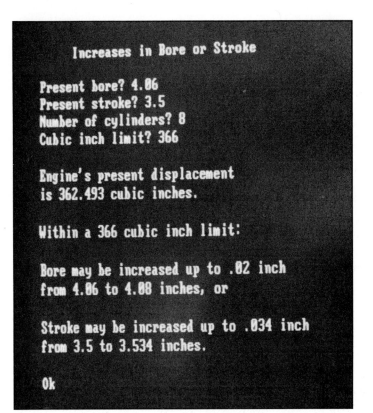

Fig. 17d. If the Ford 351 is overbored 0.060 inch, it will be up to 362.493 cubic inches. That doesn't leave much of a margin for a further increase in displacement within a 366 cubic-inch limit.

In line 140, P is assigned the value of *pi*/4. In 145, the present displacement is computed from the present bore and stroke. In 150 and 155, the bore is computed with the present stroke and the displacement limit, and the stroke with the present bore and displacement limit. In 160 and 165, the differences between the present and potential bores and strokes are figured.

Then lines 175 through 220 display the results, with lines 170, 185, 195, 210 and 225 inserting blank lines for readability.

Levels of Accuracy—In Chapter 1, an 8-digit calculator was used to find the displacement of the 4.0-inch bore, 3.5-inch stroke Ford V-8. The value of *pi* carried out to 8 digits is 3.1415927 and *pi*/4 is 0.7853982. The result of the calculation before rounding was 351.84838 cubic inches.

GW-BASIC can operate at either of two levels of accuracy—single-precision, which works with up to 7 digits, and double-precision, which works with up to 17 digits. In both cases, though, the final digit is not necessarily correct. The differences between the two are that single-precision uses less of the computer's memory and operates more quickly, while double-precision is more accurate.

In single-precision, *pi* is 3.141593 and *pi*/4 is 0.7853982. Why is the 7-digit version of the latter the same as the 8-digit version on a calculator? The calculator displays the zero in front of the decimal point and counts it as a digit, while the computer does not. At any rate, GW-BASIC single-precision would show the Ford engine's displacement as 351.8584 cubic inches.

In double-precision, *pi* is 3.141592653589793, *pi*/4 is 0.7853981633974483, and the Ford's displacement would be shown as 351.8583772020569 cubic inches.

Obviously, that's a far greater degree of accuracy than you need for measuring engine specs. Bore and stroke, specifically, are usually carried out to no more than thousandths, or three decimal places.

ROUNDING TO THOUSANDTHS

In the program, the BASIC function INT (for integer) is used in lines 180, 200, 205, 215 and 220 to round all the computations to thousandths. The INT function is extremely useful in mathematical programming, but isn't covered in many textbooks of BASIC, so I'll try to explain it briefly here.

INT is followed by either a number represented by a variable name or a number itself in parentheses. Its purpose is to reduce that number to an integer, or whole number, by eliminating any decimal fraction following it.

As noted above, GW-BASIC single-precision would compute the Ford's displacement as 351.8584 and, without the INT function, that's what the program would display on the monitor. It was pointed out in Chapter 1 that Ford rounds that down to 351 but, with a decimal greater than 0.5, it should be rounded up to 352.

Ford's inaccurate practice is supported by the computer, for, when you ask it to PRINT INT(351.8584), it displays 351. You can correct that by adding 0.5 to the number in parentheses. Typing PRINT INT(351.8584+.5) will result in a display of 352.

Retaining Decimals—But how do you retain some of the decimals? How do you get the program to display three decimal places instead of four? What follows may seem to be a complicated way to get rid of one decimal place, but it will demonstrate the principle, and it will reduce the decimal fraction to thousandths, no matter how many decimal places there are originally.

You multiply the original number by 1000, extract the integer, and then divide the result by 1000. Using the Ford's displacement, you multiply 351.8584 by 1000 to obtain 351858.4. With PRINT INT(351858.4+.5), you eliminate the decimal and get the integer 351858. Divide that by 1000 and you have 351.858, the integer with three decimal places, which is what you wanted.

Here's an example of how the computer can simplify what seems to be complicated. All that can be combined in a single program line statement. Using D for displacement, it becomes PRINT INT(D*1000+.5)/1000, and that's the form used with D1 in line 180, B3 in 200, B2 in 205, S3 in 215, and S2 in 220.

If you wanted the result to show 1 decimal point, or tenths, you would use 10 in the same way I did 1000 for 3 decimal points. If you wanted 2 decimal points, or hundredths, you would use 100.

COMMODORE 64 VERSION

The Commodore 64 also uses a version of BASIC developed by Microsoft. It's similar to GW-BASIC, but not identical.

The only change that must be made in the program as written in GW-BASIC to get it to run on a Commodore 64 is in line 105, where CLS needs to be replaced with PRINT CHR$(147). However, there are some other differences to note and other changes which can be made. A 64 can accommodate a disk file name of up to 16 characters, not just 8, so you can use something a bit more descriptive for that purpose, such as BORESTROKE64. On the other hand, the 64 requires variable names of no more than 2 characters. The GW-BASIC program was written with that in mind, so no changes are needed. The program was also written to fit a 40-column video monitor, like that of a 64.

The 64 has a *pi* key and, in line 140, you can take advantage of that by typing P=*pi*/4, rather than entering the number. The Commodore computer also works with numbers out to 9 digits, not just 7.

However, the fundamental structure of the program is the same in both GW-BASIC and Commodore 64 BASIC, and is readily transportable from one system to the other.

OTHER PROGRAMMING IDEAS

The idea is also readily adaptable to other auto math formulas. Consider the possibilities of writing BASIC programs using, say, the formulas for weight distribution in Chapter 7, center of gravity in Chapter 8, shift points in Chapter 10, or blood alcohol concentration in Chapter 16.

The structure is simple enough. To reiterate, the program asks for the known factors in a formula, processes them in a BASIC version of the formula to find the unknowns, and then displays the results on the computer monitor.

That's all there is to it. With that knowledge added to the example provided here in the program for increases in bore or stroke, you should be ready to do some automotive programming of your own. Good luck and happy hackin'!

Table 17

FORMULAS FOR AUTO MATH ON YOUR COMPUTER

P=.7853982

D=P*B^2*S*C

S=D/(P*B^2*C)

B=SQR(D/(P*S*C))

VARIABLES

P = *pi*/4

D = displacement

B = bore

S = stroke

C = number of cylinders

Appendices

Appendix A

Selected Conversion Factors

The following is a list of conversion factors for U.S. and S.I. units of measure that could be of use to the automobile enthusiast. Also included are factors for two widely used British measures: British thermal units, or Btus, and imperial gallons. As is true elsewhere in the book, abbreviations have been kept to a minimum and figures are carried out to a maximum of eight digits.

Factors for converting between two specific units of measure are reciprocals, i.e., if multiplied together, they have a product of 1.0. For example, the factor for converting from gallons to quarts is 4.0, while the factor for converting from quarts to gallons is 0.25. Multiplied together, 4.0 and 0.25 equal 1.0.

So, if you know the factor for converting in one direction but not for the other, you can find the latter by dividing the known factor into 1.0.

As a case in point, the list includes a factor of 42 for multiplying a number of barrels of oil to find the equivalent in gallons. However, no factor is given for multiplying gallons to find barrels, because that isn't a conversion many people would need to make. The 42-gallon barrel is an arbitrary unit of measure of oil used in international commerce and not a real size of container.

But suppose you did want to know how many barrels a given number of gallons of oil would equal. You can find the reciprocal of 42 by dividing it into 1.0, or 1/42, which would be 0.0238095. On a scientific calculator, you could enter 42 and then press the reciprocal key, marked 1/x.

Another way would be to divide 42 into gallons to convert to barrels. Mathematically, switching from multiplication in one direction to division in the other is the same thing as switching from multiplying by one conversion factor to multiplying by its reciprocal.

When one factor is simpler than its reciprocal, you can save time and effort by knowing when to divide instead of to multiply. In our example, it would obviously be quicker and easier to divide the gallons by 42 than to multiply the barrels by 0.0238095.

In the case of very large numbers, the reciprocal may not have enough significant digits within an eight-digit limit to be of much value. As an example, take the factor for converting from kilowatt-hours to joules, which is 3600000.0. On an eight-digit calculator, the reciprocal for converting from joules to kilowatt-hours would be 0.0000003. But, if you enter 0.0000003 and press the 1/x key, you'll get a reciprocal of 3333333.3, an error of more than 7.4 percent. The reciprocal has been rounded to only one significant digit, and that's not enough for an accurate conversion back to the original factor, nor is it enough for accurate calculations. Consequently, reciprocals with only one or two significant digits are not included in the list.

If you had to convert joules to kilowatt-hours, you'd get more accurate results by first converting the joules to kilojoules (multiplying the joules by 0.001 or dividing by 1000) and then converting the kilojoules to kilowatt-hours (multiplying the kilojoules by 0.0002778 or dividing by 3600).

Another way to deal with very large or very small numbers is to use scientific notation, a form of mathematical shorthand that eliminates the need for a large number of digits. A scientific calculator will have a key marked either EXP (for exponent) or EE (for exponent entry) for using scientific notation, and instructions for working with it will be found in the calculator's manual.

In the list, where the terms gallons, miles and ounces are used without qualification, they mean U.S. gallons, statute miles, and avoirdupois ounces, respectively. Similarly, horsepower and torque mean SAE horsepower and torque, while water means fresh water.

Factors involving water are as measured at 4.0 degrees Celsius or 39.2 degrees Fahrenheit. That's the temperature of water at its maximum density, which serves as the international standard for measuring the relative density or specific gravity of other liquids.

Finally, factors which are exact figures are indicated by an asterisk (*).

TO CONVERT FROM:	MULTIPLY BY:
atmospheres to bars	1.01325*
atmospheres to inches of mercury	29.921256
atmospheres to inches of water	406.80172
atmospheres to kilograms per square centimeter	1.0332275
atmospheres to kilopascals	101.325*
atmospheres to millibars	1013.25*
atmospheres to pounds per square inch	14.695949
barrels, non-oil liquid, to cubic feet	4.2109376
barrels, non-oil liquid, to cubic meters	0.1192405
barrels, non-oil liquid, to gallons	31.5*
barrels, non-oil liquid, to liters	119.24047
barrels, oil, to cubic feet	5.6145833
barrels, oil, to cubic meters	0.1589873
barrels, oil, to gallons	42.0*
barrels, oil, to liters	158.98729
bars to atmospheres	0.9869233
bars to inches of mercury	29.529983
bars to inches of water	401.48716
bars to kilograms per square centimeter	1.0197162
bars to kilopascals	100.0*
bars to millibars	1000.0*
bars to pounds per square foot	2088.5434
bars to pounds per square inch	14.503774
Btus to calories	251.99576
Btus to horsepower-hours	0.000393
Btus to joules	1055.0559
Btus to kilogram-meters	107.58576
Btus to kilowatt-hours	0.0002931
Btus to kilojoules	1.0550559
Btus to pounds-feet	778.16927
Btus to watt hours	0.2930711
Btus per gallon to megajoules per cubic meter	0.279
Btus per gallon to megajoules per liter	0.000279
Btus per minute to horsepower	0.0235809
Btus per minute to kilowatts	0.0175843
Btus per pound to joules per kilogram	2326.*
Btus per pound to kilojoules per kilogram	2.32*

TO CONVERT FROM:	MULTIPLY BY:
Btus per pound to megajoules per kilogram	0.002326*

calories to Btus	0.0039683
calories to joules	4.1868*
calories to kilogram-meters	0.4269348
calories to pounds-feet	3.0880252
calories to watt hours	0.001163

centiliters to deciliters	0.1*
centiliters to liters	0.01*

centimeters to feet	0.0328084
centimeters to hands	0.0984252
centimeters to inches	0.3937008
centimeters to meters	0.01*
centimeters to microns	10000.0*
centimeters to millimeters	10.0*
centimeters to mils	393.70079
centimeters to yards	0.0109361

centimeters per second to feet per second	0.0328084
centimeters per second to kilometers per hour	0.036*
centimeters per second to miles per hour	0.0223694

centimeters per second per second to feet per second per second	0.0438084
centimeters per second per second to g	0.0010197
centimeters per second per second to meters per second persecond	0.01*

circles to circumferences	1.0*

circumferences to circles	1.0*
circumferences to degrees	360.0*
circumferences to grades	400.0*
circumferences to minutes	21600.0*
circumferences to quadrants	4.0*
circumferences to radians	6.2831853
circumferences to seconds	129600.0*

cubic centimeters to cubic inches	0.0610237
cubic centimeters to cubic meters	0.000001*
cubic centimeters to gallons	0.0002642
cubic centimeters to liters	0.001*
cubic centimeters to milliliters	1.0*
cubic centimeters to ounces, fluid	0.03381
cubic centimeters to pints	0.0021134

TO CONVERT FROM:	MULTIPLY BY:
cubic centimeters to quarts	0.0010567
cubic feet to cubic centimeters	28316.847
cubic feet to cubic inches	1728.0*
cubic feet to cubic meters	0.0283168
cubic feet to cubic yards	0.037037
cubic feet to gallons	7.4805195
cubic feet to liters	28.316866
cubic feet to ounces, fluid	957.50649
cubic feet to pints	59.844156
cubic feet to quarts	29.922078
cubic feet, water, to pounds	62.424215
cubic inches to cubic centimeters	16.387064
cubic inches to cubic feet	0.0005787
cubic inches to gallons	0.004329
cubic inches to liters	0.0163871
cubic inches to ounces, fluid	0.5541126
cubic inches to pints	0.034632
cubic inches to quarts	0.017316
cubic inches, water, to pounds	0.0361251
cubic meters to cubic centimeters	1000000.0*
cubic meters to cubic feet	35.314667
cubic meters to cubic yards	1.3079506
cubic meters to gallons	264.17205
cubic meters to liters	1000.0*
cubic meters, water, to kilograms	999.94004
cubic meters, water, to pounds	2204.4903
cubic yards to cubic feet	27.0*
cubic yards to cubic meters	0.7645549
cubic yards to gallons	201.97403
cubic yards to liters	764.55486
deciliters to centiliters	10.0*
deciliters to liters	0.1*
degrees to circumferences	0.0027778
degrees to grades	1.1111111
degrees to minutes	60.0*
degrees to quadrants	0.0111111

TO CONVERT FROM:	MULTIPLY BY:
degrees to radians	0.0174533
degrees to seconds	3600.0*
fathoms to feet	6.0*
fathoms to meters	1.8288*
fathoms to yards	2.0*
feet to centimeters	30.48*
feet to fathoms	0.1666667
feet to furlongs	0.0015152
feet to inches	12.0*
feet to hands	3.0*
feet to kilometers	0.0003048
feet to meters	0.3048*
feet to mils	12000.0*
feet to yards	0.3333333
feet per second to centimeters per second	30.48*
feet per second to feet per minute	60.0*
feet per second to kilometers per hour	1.09728*
feet per second to knots	0.5924838
feet per second to meters per second	0.3048*
feet per second to miles per hour	0.6818182
feet per second per second to centimeters per second per second	30.48*
feet per second per second to g	0.0310809
feet per second per second to meters per second per second	0.3048*
furlongs to feet	660.0*
furlongs to meters	201.168*
furlongs to miles	0.125*
furlongs to yards	220.0*
g to centimeters per second per second	980.665*
g to feet per second per second	32.174049
g to meters per second per second	9.80665*
gallons to cubic centimeters	3785.4118
gallons to cubic feet	0.1336806
gallons to cubic inches	231.0*
gallons to cubic meters	0.0037854
gallons to cubic yards	0.0049511
gallons to liters	3.7854118
gallons to ounces, fluid	128.0*

TO CONVERT FROM:	MULTIPLY BY:
gallons to pints	8.0*
gallons to quarts	4.0*
gallons, acetone, to pounds	6.6
gallons, castor oil, to pounds	8.1
gallons, ethanol (ethyl alcohol), to pounds	7.6
gallons, ether, to pounds	6.2
gallons, gasoline, to pounds	6.0
gallons, kerosene, to pounds	6.6
gallons, liquid propane, to pounds	4.25
gallons, methanol (methyl alcohol), to pounds	6.7
gallons, naptha, to pounds	5.6
gallons, nitromethane, to pounds	9.4
gallons, oil (crude), to pounds	7.5
gallons, oil (refined), to pounds	7.0
gallons, turpentine, to pounds	7.3
gallons, water, to pounds	8.3449037
gallons, water (sea), to pounds	8.6
gallons, imperial, to liters	4.54609 *
gallons, imperial, to gallons, U.S.	1.2009499
gallons, U.S., to gallons, imperial	0.8326742
gallons per horsepower-hour to liters per kilowatt-hour	0.0793181
gons to grades	1.0*
grades to circumferences	0.0025*
grades to degrees	0.9*
grades to gons	1.0*
grades to minutes	54.0*
grades to radians	0.015708
grades to quadrants	0.01*
grades to seconds	3240.0*
grains to grams	0.0647989
grains to milligrams	64.798911
grains to ounces	0.0022857
grams to grains	15.432358
grams to kilograms	0.001*
grams to milligrams	1000.0*
grams to ounces	0.035274
grams to pounds	0.0022046

TO CONVERT FROM:	MULTIPLY BY:
grams per centimeter to kilograms per meter	10.0*

grams per cubic centimeter to pounds per cubic foot	62.42796
grams per cubic centimeter to pounds per cubic inch	0.0361273
grams per cubic centimeter to pounds per gallon	8.3454044
grams per cubic centimeter to kilograms per cubic meter	1000.0*

grams per kilowatt-hour to pounds per horsepower-hour	0.001644

hands to centimeters	10.16*
hands to feet	0.3333333
hands to inches	4.0*

horsepower to pounds-feet per minute	33000.0*
horsepower to pounds-feet per second	550.0*

horsepower, metric, to horsepower, SAE	0.9863201
horsepower, metric, to kilogram-meters per second	75.0*
horsepower, metric, to kilowatts	0.7354988

horsepower, SAE, to horsepower, metric	1.0138697
horsepower, SAE, to kilowatts	0.7456999

horsepower-hours to Btus	2544.4336
horsepower-hours to calories	641186.48
horsepower-hours to kilojoules	2684.52*
horsepower-hours to kilowatt-hours	0.7456999
horsepower-hours to megajoules	2.68452*

inches to centimeters	2.54*
inches to feet	0.0833333
inches to hands	0.25*
inches to meters	0.0254*
inches to microns	25400.0*
inches to millimeters	25.4*
inches to mils	1000.0*
inches to yards	0.0277778

inches of mercury to atmospheres	0.0334211
inches of mercury to bars	0.0338639
inches of mercury to inches of water	13.595915
inches of mercury to kilograms per square centimeter	0.0345316
inches of mercury to kilopascals	3.3863886
inches of mercury to millibars	33.863886
inches of mercury to pounds per square foot	70.726197

TO CONVERT FROM:	MULTIPLY BY:
inches of mercury to pounds per square inch	0.4911541
inches of water to atmospheres	0.0024582
inches of water to bars	0.0024907
inches of water to inches of mercury	0.0735515
inches of water to kilograms per square centimeter	0.0025398
inches of water to kilograms per square meter	25.398476
inches of water to kilopascals	0.249074
inches of water to millibars	2.4907397
inches of water to pounds per square foot	5.2020179
inches of water to pounds per square inch	0.0361251
joules to Btus	0.0009478
joules to calories	0.2388459
joules to kilogram-meters	0.1019716
joules to kilojoules	0.001*
joules to megajoules	0.000001*
joules to newton-meters	1.0*
joules to ounces-inches	141.61193
joules to pounds-feet	0.7375622
joules to pounds-inches	8.8507457
joules to watt-hours	0.0002778
joules to watt-seconds	1.0*
joules per gram to kilojoules per kilogram	1.0*
joules per kilogram to Btus per pound	0.0004299
kilograms to grams	1000.0*
kilograms to newtons	9.80665*
kilograms to pounds	2.2046224
kilograms to tons, long	0.0009842
kilograms to tons, short	0.0011023
kilograms, water, to liters	1.00006
kilograms per cubic meter to grams per cubic centimeter	0.001*
kilograms per cubic meter to pounds per cubic foot	0.062428
kilograms per cubic meter to pounds per gallon	0.0083454
kilograms per kilowatt-hour to pounds per horsepower-hour	1.6439879
kilograms per liter to pounds per gallon	8.3454064
kilograms per meter to grams per centimeter	0.1
kilograms per meter to pounds per foot	0.671969

TO CONVERT FROM:	MULTIPLY BY:
kilograms per meter to pounds per inch...0.0559974	
kilograms per square centimeter to atmospheres.......................................0.9678411	
kilograms per square centimeter to bars 0.980665*	
kilograms per square centimeter to inches of mercury............................28.959021	
kilograms per square centimeter to inches of water 393.7244	
kilograms per square centimeter to kilograms per square meter.........10000.0*	
kilograms per square centimeter to kilopascals.....................................98.0665*	
kilograms per square centimeter to millibars..980.665*	
kilograms per square centimeter to pounds per square foot................2048.1614	
kilograms per square centimeter to pounds per square inch 14.223343	
kilograms per square meter to kilograms per square centimeter...............0.0001*	
kilograms per square meter to pounds per square foot 0.2048161	
kilogram-meters to Btus...0.0092949	
kilogram-meters to calories 2.3422781	
kilogram-meters to joules 9.80665*	
kilogram-meters to pounds-feet 7.2330139	
kilogram-meters to watt-hours 0.0027241	
kilojoules to Btus...0.9478171	
kilojoules to horsepower-hours...0.0003725	
kilojoules to joules 1000.0*	
kilojoules to kilowatt-hours 0.0002778	
kilojoules to megajoules 0.001*	
kilojoules per kilogram to Btus per pound 0.4299226	
kilojoules per kilogram to joules per gram 1.0	
kilometers to meters 1000.0	
kilometers to miles, nautical 0.5399568	
kilometers to miles, statute 0.6213712	
kilometers per hour to centimeters per second 27.777778	
kilometers per hour to knots 0.5399568	
kilometers per hour to meters per second...0.2777778	
kilometers per hour to miles per hour 0.6213712	
kilometers per liter to miles per gallon, imperial 2.8248094	
kilometers per liter to miles per gallon, U.S. 2.3521459	
kilopascals to atmospheres 0.0098692	
kilopascals to bars...0.01*	
kilopascals to inches of mercury 0.2952998	

TO CONVERT FROM:	MULTIPLY BY:
kilopascals to inches of water	4.0148716
kilopascals to kilograms per square centimeter	0.0101972
kilopascals to millibars	10.0*
kilopascals to pascals	0.001*
kilopascals to pounds per square foot	20.885434
kilopascals to pounds per square inch	0.1450377
kilowatts to Btus per minute	56.869027
kilowatts to horsepower, metric	1.3596216
kilowatts to horsepower, SAE	1.3410221
kilowatts to newton-meters per second	1000.0*
kilowatt-hours to Btus	3412.1416
kilowatt-hours to horsepower-hours	1.3410221
kilowatt-hours to joules	3600000.0*
kilowatt-hours to kilojoules	3600.0*
kilowatt-hours to megajoules	3.6*
kilowatt-hours to watt-hours	1000.0*
knots to feet per minute	101.26859
knots to feet per second	1.6878099
knots to kilometers per hour	1.852*
knots to miles per hour	1.1507794
liters to centiliters	100.0*
liters to cubic centimeters	1000.0*
liters to cubic feet	0.0353147
liters to cubic inches	61.023744
liters to cubic yards	0.0013079
liters to deciliters	10.0*
liters to gallons, imperial	0.2199692
liters to gallons, U.S.	0.2641721
liters to milliliters	1000.0*
liters to ounces, fluid	33.814023
liters to pints	2.1133764
liters to quarts	0.879877
liters to quarts	1.0566882
liters, water, to kilograms	0.99994
liters, water, to pounds	2.2044903
liters per kilowatt-hour to gallons per horsepower-hour	12.607459
liters per kilowatt-hour to quarts per horsepower-hour	3.1518648
liters per kilowatt-hour to pints per horsepower-hour	1.5759324

TO CONVERT FROM:	MULTIPLY BY:

megajoules to joules	1000000.0*
megajoules to kilojoules	1000.0*
megajoules to kilowatt-hours	0.2777778
megajoules to horsepower-hours	0.3725061

megajoules per kilogram to Btus per pound	429.92261

meters to fathoms	0.5468067
meters to feet	3.2808399
meters to furlongs	0.004971
meters to inches	39.370079
meters to kilometers	0.001
meters to mils	39370.079
meters to yards	1.0936133

meters per second to kilometers per hour	3.6*
meters per second to feet per second	3.2808399

meters per second per second to centimeters per second per second	100.0
meters per second per second to feet per second per second	3.2808399
meters per second per second to g	0.1019716

microinches to micrometers	0.0254*

micrometers to microinches	39.370079
micrometers to microns	1.0*

microns to centimeters	0.0001*
microns to inches	0.0000394
microns to microinches	39.370079
microns to micrometers	1.0*
microns to millimeters	0.001*
microns to mils	0.0393701

miles, nautical, to feet	6076.1155
miles, nautical, to kilometers	1.852*
miles, nautical, to miles, statute	1.1597794
miles, nautical, to yards	2025.3718

miles, statute, to feet	5280.0*
miles, statute, to furlongs	8.0*
miles, statute, to kilometers	1.609344*
miles, statute, to miles, nautical	0.8689762
miles, statute, to yards	1760.0*

TO CONVERT FROM:	MULTIPLY BY:
miles per gallon, imperial, to kilometers per liter	0.3540062
miles per gallon, imperial, to miles per gallon, U.S.	0.8326742
miles per gallon, U.S., to kilometers per liter	0.4251437
miles per gallon, U.S., to miles per gallon, imperial	1.2009499
miles per hour to centimeters per second	44.704*
miles per hour to feet per minute	88.0*
miles per hour to feet per second	1.4666667
miles per hour to kilometers per hour	1.609344*
miles per hour to knots	0.8689762
millibars to atmospheres	0.0009869
millibars to bars	0.001*
millibars to inches of mercury	0.02953
millibars to inches of water	0.4014872
millibars to kilograms per square centimeter	0.0010197
millibars to kilopascals	0.1*
millibars to millimeters of mercury	0.7500617
millibars to pascals	100.0*
millibars to pounds per square foot	2.0885434
millibars to pounds per square inch	0.0145038
milligrams to grains	0.0154324
milligrams to grams	0.001*
milligrams to ounces	0.0000353
milliliters to cubic centimeters	1.0*
milliliters to liters	0.001*
milliliters to ounces, fluid	0.033814
millimeters to centimeters	0.1*
millimeters to inches	0.0393701
millimeters to meters	0.001*
millimeters to microns	1000.0*
millimeters to mils	39.370079
mils to centimeters	0.00254*
mils to feet	0.0000833
mils to inches	0.001*
mils to meters	0.0000254*
mils to microns	25.4*
mils to millimeters	0.0254*
mils to yards	0.0002778

TO CONVERT FROM:	MULTIPLY BY:
minutes to degrees..	0.0166667
minutes to grades	0.0185185
minutes to radians	0.0002909
minutes to quadrants	0.0001852
minutes to seconds	60.0*
newtons to kilograms	0.1019716
newtons to ounces	3.5969431
newtons to pounds	0.2248089
newtons per meter to newtons per millimeter	0.001*
newtons per meter to pounds per foot	0.0685218
newtons per millimeter to newtons per meter	1000.0*
newtons per millimeter to pounds per foot...............................	68.52178
newtons per millimeter to pounds per inch	5.7101471
newtons per square meter to pascals......................................	1.0*
newton-meters to joules	1.0*
newton-meters to pounds-feet..	0.7375622
ounces to grains...	437.5*
ounces to grams...	28.35*
ounces to kilograms	0.0283495
ounces to milligrams	28349.523
ounces to newtons	0.2780139
ounces to pounds	0.0625*
ounces, fluid, to cubic feet	0.0010444
ounces, fluid, to cubic inches	1.8046875
ounces, fluid, to gallons..	0.0078125*
ounces, fluid, to milliliters..	29.573529
ounces, fluid, to pints	0.0625*
ounces, fluid, to quarts	0.03125*
ounces-inches to joules or newton-meters	0.0070616
ounces-inches to pounds-feet	0.0052083
ounces-inches to pounds-inches	0.0625*
pascals to bars..	0.00001*
pascals to inches of mercury	0.0002953
pascals to inches of water	0.0040149
pascals to kilograms per square centimeter.............................	0.0000102
pascals to kilopascals	1000.0*

TO CONVERT FROM:	MULTIPLY BY:
pascals to millibars	0.01*
pascals to newtons per square meter..	1.0*
pascals to pounds per square foot	208.85434
pints to cubic centimeters..	473.17647
pints to cubic feet	0.0167101
pints to cubic inches..	28.875
pints to gallons..	0.125*
pints to liters	0.4731765
pints to ounces, fluid	16.0*
pints to quarts	0.5*
pints per horsepower-hour to liters per kilowatt-hour	0.634545*
pounds to grains	7000.0*
pounds to grams	453.6
pounds to kilograms..	0.4535924
pounds to newtons..	4.4482217
pounds to ounces	16.0*
pounds to tons, metric	0.0004536
pounds, water, to cubic feet..	0.0160194
pounds, water, to cubic inches	27.681566
pounds, water, to cubic meters	0.0004536
pounds, water, to gallons	0.1198336
pounds, water, to liters	0.4536196
pounds per cubic foot to grams per cubic centimeter...................	0.0160185
pounds per cubic foot to pounds per cubic inch	0.0005787
pounds per cubic foot to kilograms per cubic meter	16.018464
pounds per cubic foot to pounds per gallon	0.1336806
pounds per cubic inch to grams per cubic centimeter	27.679905
pounds per cubic inch to pounds per cubic foot	1728.0*
pounds per cubic inch to kilograms per cubic meter....................	27679.905
pounds per cubic inch to pounds per gallon................................	231.0*
pounds per foot to kilograms per meter	1.488164
pounds per foot to newtons per meter	14.5939
pounds per foot to newtons per millimeter..................................	0.0145939
pounds per foot to pounds per inch	12.0
pounds per gallon to grams per cubic centimeter........................	0.1198264
pounds per gallon to pounds per cubic foot................................	7.4805195
pounds per gallon to kilograms per cubic meter	119.82643

pounds per gallon to kilograms per liter	0.1198264
pounds per gallon to pounds per cubic inch	0.004329
pounds per horsepower-hour to grams per kilowatt-hour	608.2774
pounds per horsepower-hour to grams per megajoule	168.9659
pounds per horsepower-hour to kilograms per kilowatt-hour	0.6082774
pounds per inch to kilograms per meter	17.857968
pounds per inch to newtons per millimeter	0.1751268
pounds per inch to ounces per inch	16.0
pounds per inch to pounds per foot	0.0833333
pounds per square foot to atmospheres	0.0004725
pounds per square foot to bars	0.0004788
pounds per square feet to inches of mercury	0.014139
pounds per square foot to inches of water	0.1922331
pounds per square foot to kilograms per square centimeter	0.0004882
pounds per square foot to kilograms per square meter	4.8824277
pounds per square foot to kilopascals	0.0478803
pounds per square foot to millibars	0.4788026
pounds per square feet to pascals	47.880259
pounds per square foot to pounds per square inch	0.006944
pounds per square inch to atmospheres	0.068046
pounds per square inch to bars	0.0689476
pounds per square inch to inches of mercury	2.0360207
pounds per square inch to inches of water	27.681566
pounds per square inch to pounds per square foot	144.0*
pounds per square inch to kilograms per square centimeter	0.070307
pounds per square inch to kilopascals	6.8947574
pounds per square inch to millibars	68.947574
pounds-feet to Btus	0.0012851
pounds-feet to calories	0.3238316
pounds-feet to joules or newton-meters	1.3558179
pounds-feet to kilogram-meters	0.138255
pounds-feet to ounces-inches	192.0*
pounds-feet to pounds-inches	12.0*
pounds-feet to watt-hours	0.0003766
pounds-feet per minute to horsepower	0.0000303
pounds-inches to joules or newton-meters	0.1129848
pounds-inches to ounces-inches	16.0*

TO CONVERT FROM:	MULTIPLY BY:
pounds-inches to pounds-feet	0.0833333
quadrants to circumferences	4.0*
quadrants to degrees	90.0*
quadrants to grades	100.0*
quadrants to minutes	5400.0*
quadrants to radians	1.5707963
quadrants to seconds	3240000.0*
quarts to cubic centimeters	946.35295
quarts to cubic feet	0.0334201
quarts to cubic inches	57.75
quarts to gallons	0.25*
quarts to liters	0.9463529
quarts to ounces, fluid	32.0*
quarts to pints	2.0*
quarts per horsepower-hour to liters per kilowatt-hour	0.3172725
radians to circumferences	0.0159154
radians to degrees	57.295779
radians to grades	63.661978
radians to minutes	3437.7468
radians to seconds	206264.81
seconds to degrees	0.0002778
seconds to grades	0.0003086
seconds to minutes	0.0166667
seconds to radians	0.0000049
square centimeters to square feet	0.0010764
square centimeters to square inches	0.1550003
square centimeters to square yards	0.0001196
square feet to square centimeters	929.0304
square feet to square inches	144.0*
square feet to square meters	0.092903
square feet to square yards	0.1111111
square inches to square centimeters	6.4516*
square inches to square feet	0.0069444
square inches to square meters	0.0006452
square inches to square millimeters	645.16*
square inches to square yards	0.0007716

TO CONVERT FROM:	MULTIPLY BY:
square meters to square feet	10.76391
square meters to square inches	1550.0031
square meters to square yards	1.19599
square yards to square feet	9.0*
square yards to square inches	1296.0*
square yards to square meters	0.8361274
tons, long, to kilograms	1016.0469
tons, long, to pounds	2240.0*
tons, long, to tons, short	1.12*
tons, metric, to kilograms	1000.0*
tons, metric, to pounds	2204.6226
tons, metric, to tons, short	1.1023113
tons, short, to kilograms	907.18475
tons, short, to pounds	2000.0*
tons, short, to tons, long	0.8928571
tons, short, to tons, metric	0.9071847
watt-hours to Btus	3.4121416
watt-hours to calories	859.84523
watt-hours to joules	3600.0*
watt-hours to kilogram-meters	367.09784
watt-hours to kilowatt-hours	0.001*
watt-hours to pounds-feet	2655.2237
yards to centimeters	91.44*
yards to fathoms	0.5*
yards to feet	3.0*
yards to furlongs	0.0045455
yards to inches	36.0*
yards to meters	0.9144*
yards to mils	36000.0*

*Exact figure

Appendix B

Bibliography

Adler, U., editor-in-chief. *Automotive Handbook*, 2nd English edition. Stuttgart, Germany: Robert Bosch GmbH, 1986.

Alston, Chris. *Drag Race Chassis Tuning Manual.* Sacramento, California: Alston Industries, 1985.

Anand, Dev. "Dyno Secrets," *Car Craft Annual* 1988. Los Angeles: Petersen Publishing Company, 1988, pp. 167-169.

"Street Machine Reference Guide," *Car Craft,* Vol. 38, No. 2, February, 1990, pp. 67-68, 73-74.

Anderson, Edwin P. *Gas Engine Manual,* 2nd edition revised by Ted Pipe. Indianapolis: Theodore Audel and Company, 1977.

Auth, Joanne Buhl. *Deskbook of Math Formulas and Tables.* New York: Van Nostrand Reinhold Company, 1985.

Bell, A. Graham. *Performance Tuning in Theory and Practice: Four Strokes.* Somerset, England: Haynes Publishing Group, 1981.

Bird, J.O. Newnes. *Engineering Science Pocket Book.* London: William Heinemann Limited, 1987.

Bishop, Owen. *Yardsticks of the Universe.* New York: Peter Bedrick Books, 1982.

Blocksma, Mary. *Reading the Numbers: A Survival Guide to the Measurements, Numbers and Sizes Encountered in Everyday Life.* New York: Penguin Books, 1989.

Brianza, David. *Beginning Technical Mathematics Made Easy.* Blue Ridge Summit, Pennsylvania: TAB Books, 1990.

Campbell, Colin. *The Sports Car: Its Design and Performance,* 4th edition. Cambridge, Massachussets: Robert Bentley, 1978.

The Sports Car Engine: Its Tuning and Modification. Cambridge, Massachusetts: Robert Bentley, 1964.

Carmichael, Robert D., and Edwin R. Smith. *Mathematical Tables and Formulas.* New York: Dover Publications, 1962.

Chevrolet Power, 4th edition. Warren, Michigan: Chevrolet Motor Division, General Motors Corporation, 1980.

Christy, John, editor. *Supertuning.* New York: New American Library, 1966.

Chrysler Kit Car. Catalog SP11. Detroit: Chrysler Corporation, 1977.

Corn, Juliana, and Tony Behr. *Technical Mathematics through Applications*. Philadelphia: Saunders College Publishing, 1982.

Csere, Csaba. "Torque of the Town," *Car and Driver*, Vol..36, No. 1, July 1990, pp. 24-25.

Emiliani, Cesare. *The Scientific Companion: Exploring the Physical World with Facts, Figures, and Formulas*. New York: John Wiley and Sons, 1988.

Estes, Bill. *The RV Handbook*. Agoura, California: Trailer Life Books, 1991.

"Weigh Your RV," *RV Do It Yourself*. Calabasas, California: Trailer Life Books, 1975, pp. 26-29

"Figure It Out!" *Drag Racing USA*, Vol. 6, No. 9, June 1970, pp. 45-47, 60-61, 70.

Fitch, James W. *Motor Truck Engineering Handbook,* 3rd edition. Anacortes, Washington: Published by the author, 1984.

Flammang, James. *Understanding Automotive Specifications and Data*. Blue Ridge Summit, Pennsylvania: TAB Books, 1986.

Francisco, Don. "Math and Formulas for Hot Rodders," *Hot Rod Magazine* Yearbook Number One. Los Angeles: Petersen Publishing Company, 1961, pp. 35-39

Greenberg, Leon A. *Alco-Calculator: An Educational Instrument*. Piscataway, New Jersey: Rutgers University Center of Alcohol Studies, 1983.

Guido, Raymond. *Calculating with BASIC*. Milford, Connecticut: Scelbi Publications, 1979.

Hale, Patrick. "Drag Strip Dyno," *The First 60 Feet: A Newsletter for the Quarter User's Group*, July 1987.

Quarter, Jr. Computer software. Phoenix, Arizona: Racing Systems Analysis, 1987. Apple, Commodore and IBM disks.

High Performance Engines. Dearborn, Michigan: Ford Motor Company, 1969.

Hills, Herbert *Impco Carburetion*. Cerritos, California: Impco Carburetion, n.d.

Hudson, Ralph G., S.B. *The Engineers' Manual,* 2nd edition. New York: John Wiley and Sons, 1979.

Huntington, Roger. *American Supercar*. Los Angeles, California: Price Stern Sloan/HPBooks, 1983.

"True Power," *Car Life*, Vol. 17, No. 4, May 1970, pp. 10-13.

Jarman, Trant. "Let's Torque," *Automobile,* Vol. 2, No. 3, June 1987, pp. 40-41.

Jennings, Gordon. *Two-Stroke Tuner's Handbook*. Los Angeles, California: Price Stern Sloan/HPBooks, 1973

Jute, Andre. Designing and Building Special Cars. London: B.T. Batsford, 1985.

Kassab, Vincent. *Technical BASIC*. Englewood Cliffs, New Jersey: Prentice-Hall, 1984.

Klein, Herbert Arthur. *The Science of Measurement: A Historical Survey*. New York: Dover Publications, 1988.

Landis, Bob. "Racer Arithmetic," *The Racer's Complete Reference Guide*. Santa Ana, California: Steve Smith Autosports, 1976, pp. 196-199.

Losee, Jim. "Guide to Formulas and Conversions," *Car Craft*, Vol. 37, No. 12, December 1989, pp. 52-55.

Ludlam, F.W. *The Elementary Theory of the Internal Combustion Engine,* 3rd edition. London and Glasgow: Blackie and Sons, 1947.

Martin, Mike. *Mopar Suspensions*. Brea, California: S-A Design Books, 1984.

McFarland, Jim. *The Great Manifold Bolt-On*. El Segundo, California: Edelbrock Corporation, 1982.

"Hot Rod Shop Series: Basic Mathematics for the Car Enthusiast," *Hot Rod Magazine*, Vol. 33, No. 1, January 1980, pp. 54-59.

"Power Theory," *Engines, Hot Rod High Performance Series*, Vol. 4, No. 1, 1987, pp. 24-26.

"Street Machine Math," *Car Craft Yearbook*, Los Angeles: Petersen Publishing Company, 1987, pp. 87-89.

Metric Conversion Tables. Woodbury, New York: Barron's Educational Series, 1976.

Moore, Claude S., Bernie L. Griffin, and Edward C. Polhamus Jr. *Applied Math for Technicians*, 2nd edition. Englewood Cliffs, New Jersey: Prentice-Hall, 1982.

Newton, K., W. Steeds and T.K. Garrett. *The Motor Vehicle*, 11th edition. London: Butterworths, 1989.

NHRA Drag Rules. Glendora, California: National Hot Rod Association, published annually.

Oddo, Frank. "Thumbs Up on Engine Life," *Popular Cars*, Vol. 8, No. 6, June 1986, p. 66.

Patterson, G.A. *Engine Thermodynamics with a Pocket Calculator*, 2nd edition. Palos Verdes Estates, California: Basic Science Press, 1983.

Pitt, Jerry. "Time for Torque," *Car Craft*, Vol. 36, No. 11, November 1988, pp. 34-37.

Puhn, Fred. *How to Make Your Car Handle*. Los Angeles, California: Price Stern Sloan/HPBooks, 1976.

Richmond, Doug. *Metrics for Mechanics*. Berkeley, California: Dos Reales Publishing, 1974.

Roe, Doug. *Rochester Carburetors*, revised edition. Los Angeles, California: Price Stern Sloan/HPBooks, 1986.

Rogowski, Stephen J. *Computers for Sea and Sky*. Morristown, New Jersey: Creative Computing Press, 1982.

SAE Handbook, 4 vols. Warrendale, Pennsylvania: Society of Automotive Engineers, published annually.

Schimizzi, Ned V. *Mastering the Metric System*. New York: New American Library, 1975.

Schofield, Miles. "Basic Engine Math," *Basic Engine Hot Rodding*. Los Angeles: Petersen Publishing Company, 1972, pp. 14-17.

Shepard, Larry S. *How to Hot Rod Small-Block Mopar Engines*. Los Angeles: HPBooks, 1989

Mopar Chassis Speed Secrets. Farmington Hills, Michigan: Chrysler Corporation Direct Connection, 1984.

Mopar Oval Track Modifications. Farmington Hills, Michigan: Chrysler Corporation Direct Connection, 1983.

"Tech Tips," *Mopar Performance News*, "Finding the Head CCs," Vol. 8, No. 3, April 1987, p. 8; "CC'ing the Block," Vol. 8, No. 4, May 1987, pp. 8-9; ".500" Down Fill Volumes," Vol. 8, No. 5, June 1987, p. 10; "Compression Ratio," Vol. 8, No. 6, July 1987, pp. 14-15; "Gearing," Vol. 10, No. 9, October 1989, p. 11.

Sherman, Don. "Pfederstarke and Other Horsepower Secrets Revealed," *Car and Driver*, Vol. 30, No. 12, June 1985, pp. 26-27.

Simanaitis, Dennis. "Technical Tidbits," *Road & Track*, Vol. 38, No. 2, October 1986, pp. 148-149.

"Technical Correspondence," *Road & Track*, "Let's Ask the Professor," Vol. 37, No. 10, June 1986, pp. 206-207; "Engine Output and Altitude," Vol. 39, No. 9, May 1988, pp. 166-168; "Calculating RPM and Speed," Vol. 39, No. 10, June 1988, pp. 158-160, 162; "Predicting Horsepower," Vol. 42, No. 2, October 1990, pp. 170-171.

Smith, Carroll. *Tune to Win*. Osceola, Wisconsin: Classic Motorbooks, 1978.

Smith, Philip H. *The Design and Tuning of Competition Engines*, 6th edition revised by David N. Wenner. Cambridge, Massachusetts: Robert Bentley, 1977.

Smith, Steve. *Advanced Race Car Suspension Development*, revised edition. Santa Ana, California: Steve Smith Autosports, 1975.

Stearn, Marshall B., Ph.D. *Drinking and Driving: Know Your Limits and Liabilities*. Sausalito, California: Park West Publishing Company, 1985.

Storer, Jay. "What Is Horsepower?" *Hot Rod Yearbook* Number 13. Los Angeles: Petersen Publishing Company, 1973, pp. 114-117.

Taborek, Jaroslav J. *Mechanics of Vehicles*. Cleveland, Ohio: Penton Publishing Company, 1957.

Taxel, I. *Conversion Factors with Metric Calculator*. Woodmere, New York: Published by the author, 1964.

Tire Inflation and Substitution/Performance Formulas. Akron, Ohio: BFGoodrich, 1986.

Titus, Rick. "The Pursuit of Control," *Hot Rod 1986 Annual*. Los Angeles: Petersen Publishing Company, 1985, pp. 146-151.

Urich, Mike, and Bill Fisher. *Holley Carburetors & Manifolds*. Los Angeles, California: Price Stern Sloan/HPBooks, 1976.

Van Valkenburgh, Paul. *Race Car Engineering and Mechanics*, 2nd edition. Seal Beach, California: Published by the author, 1986.

Von Helmolt, Ken. "Differential Equations," *4x4 Answer Book*. Canoga Park, California: Four Wheeler Publishing, 1987, pp. 69-71.

Wallace, Dave. "Chassis Tune-Up," *Hot Rod 1986 Annual*. Los Angeles: Petersen Publishing Company, 1985, pp. 88-98.

Wilson, Waddell, and Steve Smith. *Racing Engine Preparation*. Santa Ana, California: Steve Smith Autosports, 1975.

Winchell, Frank. "A Lesson in Basic Vehicular Physics," *Automobile Magazine*, Vol. 2, No. 9, December 1987, pp. 45-50.

Index

Load, 25, 116-117
Low-profile tires, 95
Lumina, 106

M

Machine, 5, 83-84, 139, 141
Manifold, 140
Mathematical, 4, 61, 77-78, 117-118, 139
Mazda, 49
Measuring, 10, 12-13, 25, 47, 59, 61, 65, 69, 118
Mercedes-Benz, 6, 49

Metabolism, 110
Metal, 15, 60, 148
Metallurgy, 18, 20
Meter, 27
Methanol, 128, 134
Metrics, 141
Miata, 49
Micrometer, 3
Microsoft, 115-116, 119
Mileage, 103-104
Miles-per-hour, 78
Millibars, 124, 130-131, 133, 135
Milliliters, 123, 125, 131, 133
Milling, 14-15
Modifications, 2, 77, 79, 117, 141
Modified, 27, 29, 38, 68, 73-74, 117
Modify, 3, 32
Mopar, 11-12, 19, 80, 140-141, 148
Multiplication, 54, 116, 123
Musclecars, 40
Mustang, 89-92, 99, 148

N

Nautical, 130, 132
Newton-meters, 130, 133

Newton, 131, 141
Newtons, 133
NHRA, 4, 6, 141
NHTSA, 109
Nissan, 5, 58
Nitromethane, 128, 134
Nostrand, 139
Numerator, 18, 90

O

Octane, 9
Off-center, 59, 64
Off-highway, 96, 99
Off-road, 95, 97
Offset, 50
Offsetting, 54
Oldsmobile, 84
Overbore, 3
Oversteer, 49

P

Physics, 142
Pints, 127, 134-135
Pistons, 1, 12-13
Plexiglass, 12
Plot, 75
Pontiac, 90
Porsche, 49
Porting, 39
Pound-foot, 75
Pour, 12
Power-to-weight, 78
Powerplant, 20, 50
Prony brake, 25
Propane, 128, 134
Proportion, 33, 49, 100, 110
PSI, 32-34
Publications, 111, 139-140

Pythagoras, 61

R

Radial, 96-97
Radians, 126, 128, 135-136
Radii, 100
Ramcharger, 98-99
Rear-drive, 48-49, 59
Rear-engined, 49
Recalibration, 96
Reciprocal, 123
Rodder, 32, 115, 148
Rodding, 18, 141
Rods, 18, 20, 40, 60
Rotational, 24
Rounding off, 2, 4, 24, 78, 118

S

SAE, 34, 129, 131, 141
Safety, 109
Science, 23, 139-141
Scientific, 2, 65, 104, 123, 140
Sexagesimal, 105
Short-stroke, 20
Simplify, 24, 28, 74, 116, 119
Simplifying, 32, 116
Skid pad, 69-70
Small-block, 2, 10, 12, 18, 20, 141, 148
Small-bore, 34
Squared, 2, 32, 61, 65
Squares, 61
Statistics, 110
Stopwatch, 84-85
Stroked, 32, 117
Strokes, 18-20, 31-32, 37, 118, 139
Subtract, 48, 54, 59, 85, 111
Subtracting, 14, 48, 54, 74, 84

HANDBOOKS

Auto Electrical Handbook: 0-89586-238-7
Auto Upholstery & Interiors: 1-55788-265-7
Brake Handbook: 0-89586-232-8
Car Builder's Handbook: 1-55788-278-9
Street Rodder's Handbook: 0-89586-369-3
Turbo Hydra-matic 350 Handbook: 0-89586-051-1
Welder's Handbook: 1-55788-264-9

BODYWORK & PAINTING

Automotive Detailing: 1-55788-288-6
Automotive Paint Handbook: 1-55788-291-6
Fiberglass & Composite Materials: 1-55788-239-8
Metal Fabricator's Handbook: 0-89586-870-9
Paint & Body Handbook: 1-55788-082-4
Sheet Metal Handbook: 0-89586-757-5

INDUCTION

Holley 4150: 0-89586-047-3
Holley Carburetors, Manifolds & Fuel Injection: 1-55788-052-2
Rochester Carburetors: 0-89586-301-4
Turbochargers: 0-89586-135-6
Weber Carburetors: 0-89586-377-4

PERFORMANCE

Aerodynamics For Racing & Performance Cars: 1-55788-267-3
Baja Bugs & Buggies: 0-89586-186-0
Big-Block Chevy Performance: 1-55788-216-9
Big Block Mopar Performance: 1-55788-302-5
Bracket Racing: 1-55788-266-5
Brake Systems: 1-55788-281-9
Camaro Performance: 1-55788-057-3
Chassis Engineering: 1-55788-055-7
Chevrolet Power: 1-55788-087-5
Ford Windsor Small-Block Performance: 1-55788-323-8
Honda/Acura Performance: 1-55788-324-6
High Performance Hardware: 1-55788-304-1
How to Build Tri-Five Chevy Trucks ('55-'57): 1-55788-285-1
How to Hot Rod Big-Block Chevys:0-912656-04-2
How to Hot Rod Small-Block Chevys:0-912656-06-9
How to Hot Rod Small-Block Mopar Engines: 0-89586-479-7
How to Hot Rod VW Engines:0-912656-03-4
How to Make Your Car Handle:0-912656-46-8
John Lingenfelter: Modifying Small-Block Chevy: 1-55788-238-X
Mustang 5.0 Projects: 1-55788-275-4

Mustang Performance ('79-'93): 1-55788-193-6
Mustang Performance 2 ('79-'93): 1-55788-202-9
1001 High Performance Tech Tips: 1-55788-199-5
Performance Ignition Systems: 1-55788-306-8
Performance Wheels & Tires: 1-55788-286-X
Race Car Engineering & Mechanics: 1-55788-064-6
Small-Block Chevy Performance: 1-55788-253-3

ENGINE REBUILDING

Engine Builder's Handbook: 1-55788-245-2
Rebuild Air-Cooled VW Engines: 0-89586-225-5
Rebuild Big-Block Chevy Engines: 0-89586-175-5
Rebuild Big-Block Ford Engines: 0-89586-070-8
Rebuild Big-Block Mopar Engines: 1-55788-190-1
Rebuild Ford V-8 Engines: 0-89586-036-8
Rebuild Small-Block Chevy Engines: 1-55788-029-8
Rebuild Small-Block Ford Engines:0-912656-89-1
Rebuild Small-Block Mopar Engines: 0-89586-128-3

RESTORATION, MAINTENANCE, REPAIR

Camaro Owner's Handbook ('67-'81): 1-55788-301-7
Camaro Restoration Handbook ('67-'81): 0-89586-375-8
Classic Car Restorer's Handbook: 1-55788-194-4
Corvette Weekend Projects ('68-'82): 1-55788-218-5
Mustang Restoration Handbook('64 1/2-'70): 0-89586-402-9
Mustang Weekend Projects ('64-'67): 1-55788-230-4
Mustang Weekend Projects 2 ('68-'70): 1-55788-256-8
Tri-Five Chevy Owner's ('55-'57): 1-55788-285-1

GENERAL REFERENCE

Auto Math:1-55788-020-4
Fabulous Funny Cars: 1-55788-069-7
Guide to GM Muscle Cars: 1-55788-003-4
Stock Cars!: 1-55788-308-4

MARINE

Big-Block Chevy Marine Performance: 1-55788-297-5

HPBOOKS ARE AVAILABLE AT BOOK AND SPECIALTY RETAILERS OR TO
ORDER CALL: 1-800-788-6262, ext. 1

HPBooks
A division of Penguin Putnam Inc.
375 Hudson Street
New York, NY 10014